HENRY C. TUCKWELL
Monash University

Stochastic Processes in the Neurosciences

SOCIETY FOR INDUSTRIAL AND APPLIED MATHEMATICS

PHILADELPHIA, PENNSYLVANIA   1989

Copyright 1989 by the Society for Industrial and Applied Mathematics

All rights reserved. No part of this book may be reproduced, stored, or transmitted in any manner without the written permission of the Publisher. For information, write the Society for Industrial and Applied Mathematics, 3600 University City Science Center, Philadelphia, Pennsylvania 19104-2688.

Typeset by The Universities Press (Belfast) Limited, Northern Ireland; Printed by Capital City Press, Montpelier, Vermont.
Second printing 1994.

Library of Congress Cataloging-in-Publication Data

Tuckwell, Henry C. (Henry Clavering), 1943 –
   Stochastic processes in the neurosciences.

   (CBMS–NSF regional conference series in applied mathematics; 56)
   Based on lectures given by the author at the CBMS–NSF regional conference in June 1986 at North Carolina State University.
     Bibliography: p.
     Includes index.
     1. Stochastic processes—Congresses. 2. Neurology—Mathematical models—Congresses. 3. Neurology—Statistical analysis—Congresses. 4. Neural transmission—Mathematical models—Congresses. I. Title. II. Series.

QP356.T78   1989   599'.0188   89-6115
ISBN 0-89871-232-7

siam. is a registered trademark.

Contents

v PREFACE

1 CHAPTER 1. Deterministic Theories and Stochastic Phenomena in Neurobiology

11 CHAPTER 2. Synaptic Transmission

15 CHAPTER 3. Early Stochastic Models for Neuronal Activity

29 CHAPTER 4. Discontinuous Markov Processes with Exponential Decay

43 CHAPTER 5. One-Dimensional Diffusion Processes

57 CHAPTER 6. Stochastic Partial Differential Equations

81 CHAPTER 7. The Statistical Analysis of Stochastic Neural Activity

91 CHAPTER 8. Channel Noise

101 CHAPTER 9. Wiener Kernel Expansions

111 CHAPTER 10. The Stochastic Activity of Neuronal Populations

117 REFERENCES

127 INDEX

To Tania

Preface

This monograph is based on lectures given by the author at the National Science Foundation–Conference Board of the Mathematical Sciences Regional Conference in June 1986 at North Carolina State University, Raleigh, North Carolina.

Chapter 1 contains introductory material in the form of a brief survey of some relevant biology and deterministic theories. Chapter 2 looks at some stochastic phenomena connected with synaptic transmission. The third chapter summarizes early neural modeling, from 1941 to 1964. There follow two chapters (Chapters 4 and 5) on one-dimensional Markov processes, both discontinuous and continuous. Chapter 6 is concerned with linear and nonlinear stochastic partial differential equations. The statistical analysis of neuronal data is then presented, followed by a chapter on channel noise. Chapter 9 deals with Wiener kernel expansions, and the final chapter is concerned with stochastic phenomena in neuronal populations.

I would like to thank Professor Charles E. Smith for organizing this meeting and the National Science Foundation for its financial support. I am grateful to several colleagues for their contributions to the material presented in this volume, in particular, Professors Davis K. Cope, Floyd B. Hanson, John B. Walsh, and Frederic Y. M. Wan. I also thank Ms. Babette Dalton and Ms. Barbara Young for the excellent manuscript preparation.

CHAPTER 1

Deterministic Theories and Stochastic Phenomena in Neurobiology

In order to make more meaningful the theory of stochastic processes that arise in connection with nervous systems, we will first collect some basic neurobiological facts and briefly review some of the major deterministic theories for the behaviors of single neurons. An explanation of various stochastic phenomena for which mathematical theories have been proposed, or for which mathematical analysis has proved fruitful, will follow. However, we will precede this discussion with a short historical overview.

1.1. A brief historical review.

It was in 1939 that the first intracellular recording of the electrical potential difference across a nerve cell membrane was made by Hodgkin and Huxley. That recording was made on the giant axon of squid and enabled the details of action potentials—brief propagating waves of voltage change—to be observed. A few years later Hodgkin and Rushton (1946) showed that the linear cable equation was successful in predicting axonal voltages at moderate levels of excitation, levels less than those that led to action potentials.

Pioneering studies on mammalian nerve cells were made by Eccles and his colleagues in 1952 and subsequent years (see, e.g., Eccles (1957)). These studies were made possible by the development of microelectrode recording techniques, and the chief cell investigated was the spinal motoneuron.

At about the same time two other significant advances were made: (1) the use of the electron microscope (for the first time it became possible to see the fine structure of nerve surfaces and the junctions between different cells); and (2) the development of a phenomenological model, consisting of a system of nonlinear differential equations for nerve membrane potential, which encompassed both subthreshold phenomena and action potential instigation and propagation.

With estimates of a neuron's resistance and capacitance available, the R-C circuit model of Lapicque (1907) was revived. However, questions arose as to the role played by the dendrites in a neuron's integrative behavior. Rall (1959) began a series of investigations into the responses of neurons with various geometries, using linear cable theory for each segment. Meanwhile, an attack on modeling a complex nerve net was made by Beurle (1956).

Neurobiology had thus become mathematized, but it was not until the 1960s that stochastic processes played a significant role. Earlier models, not being constructed with the insights obtained with microelectrodes and the electron microscope, had tended to be ad hoc and, in hindsight, inadequate. In 1964, Gerstein and Mandelbrot proposed a random walk model for the subthreshold potential of a spontaneously active nerve cell. The random walk was approximated by a Wiener process with drift, and shortly thereafter the Ornstein–Uhlenbeck Process (OUP) was employed as a more realistic model. Furthermore, at about the same time, Stein (1965) proposed a discontinuous Markov process which, like the OUP, incorporated the exponential decay of membrane potential.

The processes with decay and their generalizations presented first passage time problems, which were difficult to solve. The weak convergence of the discontinuous processes to diffusions has been established only very recently. In the last few years stochastic partial differential equations have been introduced to allow for the spatial extent of neurons, with particular reference to dendrites. Some of these have solutions that involve infinite-dimensional processes and present new and difficult problems of analysis.

Stochastic processes that arise in modeling single nerve cells and populations of them are not the only facet of this subject area. The statistical analysis of neuronal data is expected to be valuable in obtaining estimates of neurophysiological and neuroanatomical parameters and also in elucidating mechanisms involved in perception. There has also been a lot of attention focused on ionic channel noise, especially since recordings were made from single channels (Neher and Sakmann (1976)). The processes that arise in that context are continuous time Markov chains. Other areas include the following: Wiener kernel expansions, which are employed in the analysis of some complex neuronal networks; filtering problems for evoked potentials; and the spectral analysis of stationary and nonstationary portions of the electroencephalogram, which had so fascinated Wiener.

1.2. Some basic neuroanatomy and neurophysiology.

The broadest features of a mammalian central nervous system (CNS) consist of *brain* and *spinal cord*. The major components of the brain are the *cerebral hemispheres,* which are linked with the millions of nerve fibers which constitute the *corpus callosum*. In man, each hemisphere, if laid out flat (i.e., deconvoluted), ould have an area of about 1,200 square cm and would be about 3 mm thick.

The *neuron doctrine,* chiefly associated with the name of Ramón y Cajal (circa 1900), states that nervous systems (usually) are composed of discrete units—cells called *neurons*. The common estimate is that there are of order 10^{10} neurons in the cerebral cortex (the gray matter of the cerebral hemispheres). There are perhaps around 2.5×10^{10} neurons in a whole brain.

Within the cerebral cortex, neuron cell bodies tend to occur in layers; there are bands of fibers along the layers and also perpendicular to them. This structure supports the idea of a *columnar* or *modular* arrangement of the neocortex. Each column occupies about 250 square microns. There are estimated to be about four million such modules, each one containing around 2,500 neurons. There are connections between columns on the same (ipsilateral) and opposite (contralateral) sides of the brain. This means that when afferent (incoming) impulses excite one column, the activity spreads to a sequence of columns in a definite spatiotemporal pattern. Note that only 5 percent of the neocortex is believed to be concerned with sensory input or motor output. These and many more fascinating details are summarized in the wonderful book of Eccles (1984).

1.2.1. Properties of single neurons. There are many distinctive neuron morphologies, as the differences between, for example, pyramidal cells, spinal motoneurons, and Purkinje cells of the cerebellum indicate. Nevertheless, it is useful to conceptualize a paradigm nerve cell in the following manner. A branching structure composed of *dendrites* and referred to as a *dendritic tree* (or *trees* if there are several primary dendrites) originates from a relatively compact *cell body* or *soma*. A typical soma dimension is 20 to 50 microns; typical dendritic diameters range from a few to 10 microns. The dendrites and cell body represent the portions of the cell that receive incoming signals. From a soma will usually project an *axon,* which transmits action potentials to its endings, called *telodendria*; these, in turn, often make contact with other nerve cells or muscle cells.

A brief account of the *electrophysiology* of nerve cells is as follows. Penetration of a neuron's membrane with a microelectrode shows that when a cell is in the resting state, the electrical potential is about -70 mV inside relative to that of the external medium. The membrane potential V_m is, ignoring spatial variations, defined as the inside potential minus the outside potential. The latter is usually set arbitrarily at zero, so the resting membrane potential is $V_{m,R} \cong -70$ mV. The *depolarization* is defined as

$$V = V_m - V_{m,R}.$$

A cell is said to be *excited* (or depolarized) if $V > 0$ and *inhibited* (or hyperpolarized) if $V < 0$.

In many neurons, when a sufficient (*threshold*) level of excitation is reached, an *action potential* may occur. A typical threshold is about 10 mV relative to rest, but this is a great oversimplification. During the action potential a transient depolarization of perhaps up to 100 mV occurs which propagates along the axon and often in the reverse direction, back into the soma—the dendritic region.

1.2.2. Synapses. Nerve cells may be excited or inhibited by natural or controlled experimental stimuli. A natural mode of stimulation is through a neuron's *synapses*. We may define a synapse as a junction between nerve cells (or between muscle and nerve cell) such that electrical activity in one cell may influence the electrical potential in the other. It follows that synapses may be *excitatory* or *inhibitory*. Usually the cell in which the voltage change is induced is called the *postsynaptic* cell, as opposed to the other cell, which is referred to as *presynaptic*. The potential changes induced in the postsynaptic cell are called postsynaptic potentials (excitatory, EPSP; or inhibitory, IPSP) and usually consist of a transient departure from resting potential whose initial phase is relatively rapid and whose final phase consists of an exponential decay to resting level. However, it is now known that the features of a postsynaptic potential are determined by, amongst other things, the location of the synapse, assuming that recordings are being made from the cell body. Eccles (1964) contains many examples of postsynaptic potentials.

Estimates of quantities such as cell densities and synapse densities involve painstaking laboratory studies and come under the heading of stochastic neuronatomy, which is mentioned in Chapter 10. Figures as high as 100,000 synapses per neuron in cerebral or cerebellar cortex have been reported. For cat spinal motoneurons, a total number of 22,600 on average was estimated (Koziol and Tuckwell (1978)) from Conradi's (1969) and Barrett and Crill's (1974) data. Of these, 12,000 were excitatory; so we see that preventing a cell from firing is as important as making it fire. Using some data on cat spinal motoneurons, the number density of synapses per unit length as a function of distance x from the soma in microns can be estimated as

$$n(x) = 60 - 0.1x, \qquad 0 < x < 600.$$

This gives a reasonable estimate of 18,000 for the total number of synapses. Naturally, large numbers of synapses like these are very closely packed on a neuron's surface. Introductory accounts of neurophysiology are contained in Kuffler and Nicholls (1976) and Junge (1981).

1.3. Deterministic theories of the electrophysiological properties of neurons.

Voltages across nerve cell membranes come under three categories:
 (i) equilibrium voltages as exemplified by the resting potential;
 (ii) local transients usually dominated by passive membrane properties;
 (iii) action potentials.

We will summarize key facts and results in relation to these.

Equilibrium voltages. In the absence of stimulation, an equilibrium exists relative to diffusive and electrical forces. Under the assumption of a constant electric field within the membrane, the equilibrium voltage is given by the Goldman–Hodgkin–Katz formula,

$$V_m = \frac{RT}{F} \ln \left[\frac{P_K[K]_o + P_{Na}[Na]_o + P_{Cl}[Cl]_i}{P_K[K]_i + P_{Na}[Na]_i + P_{Cl}[Cl]_o} \right],$$

where $[\ ]_{i,o}$ denote intracellular and extracellular ionic concentrations; P_K, P_{Na}, and P_{Cl} are the permeability coefficients of potassium, sodium, and chloride ions, respectively; and RT/F is a constant $\cong 25.2$ mV at 20°C.

If a membrane is permeable to just one ionic species, then this reduces to the corresponding Nernst potential. For example, for potassium

$$V_K = \frac{RT}{F} \ln\left[\frac{[K]_o}{[K]_i}\right].$$

These formulas are quite accurate over a variety of conditions, even though the theories on which they are based are not the most recent.

Lapicque model. The origins of this model can be found in the work of Lapicque (1907), although the experiments that substantiate its details were not performed until many years later (Eccles (1957)). Here the neuron is considered to be a lumped circuit of an ohmic resistance and capacitance in parallel, driven by current sources representing synapses or experimental electrodes. Thus

$$C\frac{dV}{dt} + \frac{V}{R} = I,$$

where C = capacitance, R = resistance, V = voltage, and I = current. This equation is valid for $V < \theta$, where θ is a threshold (constant or varying) for action potentials. The imposition of the threshold makes the model nonlinear. This is the basis of Stein's model and the OUP model (Chapters 4 and 5).

Linear cable theory. Let $V(x, t)$ be the depolarization at a space point x cm along a nerve cylinder at time t sec. Then the equation

$$r_m c_m \frac{\partial V}{\partial t} = \frac{r_m}{r_i} \frac{\partial^2 V}{\partial x^2} - V + r_m I_A, \quad a < x < b, \quad t > 0$$

with appropriate boundary data was found to describe satisfactorily many subthreshold voltage responses in certain axons and has been employed for dendritic segments. Here r_m is the membrane resistance of unit length times unit length, c_m is the membrane capacitance per unit length, r_i = internal (axial) resistance per unit length, and I_A is the applied current density. Introducing dimensionless variables $x' = x/\lambda$, $t' = t/\tau$, where $\lambda = (r_m/r_i)^{1/2}$, $\tau = c_m r_m$ are the space and time constants, and putting $I' = kI\lambda\tau/d^{3/2}$ and then dropping all primes, we get the simpler equation

$$\frac{\partial V}{\partial t} = \frac{\partial^2 V}{\partial x^2} - V + I,$$

where d is the diameter of the nerve cylinder in cm,

$$k = \frac{2}{\pi C_m} \left(\frac{\rho_i}{\delta \rho_m}\right)^{1/2}$$

is a constant of the neuron, C_m = membrane capacitance per square cm, ρ_i is the intracellular resistivity, ρ_m is the membrane resistivity, and δ is the membrane thickness.

1.3.1. Dendritic trees and mappings to cylinders. Consider a single dendritic tree with a soma represented by a point zero. Suppose that dendritic branching occurs at electrotonic distances $x_1, x_2, \cdots, x_m$ from 0 and put $x_o = 0$. Assume that all dendritic terminals (i.e., the remotest dendritic endings) are at the same distance $l = x_{m+1}$ from 0 and there are n of them. Let us allow for two or more daughter cylinders at a branch point but insist on conservation of diameter$^{3/2}$. That is, if d_0 is the diameter of any parent cylinder and $d_1, d_2, \cdots$ are the diameters of the daughter cylinders, then we require

(1.1) $$d_0^{3/2} = d_1^{3/2} + d_2^{3/2} + \cdots.$$

On each dendritic segment the voltage or depolarization is assumed to satisfy a linear cable equation. Let there be n_i cylinders in existence between x_i and x_{i+1}, $i = 0, 1, \cdots, m$; let their corresponding depolarizations, applied current densities, and diameters be V_{ij}, I_{ij}, d_{ij}, $j = 1, 2, \cdots, n_i$. Thus

$$\frac{\partial V_{ij}}{\partial t} = \frac{\partial^2 V_{ij}}{\partial x^2} - V_{ij} + \frac{k}{d_{ij}^{3/2}} I_{ij}, \quad x_i < x < x_{i+1}, \quad t > 0.$$

Given the boundary conditions at the terminals,

$$\left[\alpha V_{mj}(x, t) + \beta \frac{\partial V_{mj}}{\partial x}(x, t) \right]_{x=l} = \gamma_j(t), \quad j = 1, 2, \cdots, n, \quad t > 0,$$

a boundary condition at the soma,

$$\left[a V_{01}(x, t) + b \frac{\partial V_{01}}{\partial x}(x, t) \right]_{x=0} = c(t), \quad t > 0,$$

and a set of initial depolarizations,

$$V_{ij}(x, 0) = v_{ij}(x), \quad j = 1, 2, \cdots, n_i, \quad x_i < x < x_{i+1},$$

we have the following result (Walsh and Tuckwell (1985)).

THEOREM 1.1. *The weighted sum of the depolarizations*

$$V(x, t) = \sum_{j=1}^{n_i} \left(\frac{d_{ij}}{d_{01}} \right)^{3/2} V_{ij}(x, t), \quad i = 0, 1, \cdots, m, \quad x_i < x < x_{i+1},$$

is the solution of the cable equation

(1.2) $$\frac{\partial V}{\partial t} = \frac{\partial^2 V}{\partial x^2} - V + \frac{k}{d_{01}^{3/2}} I(x, t), \quad 0 < x < l, \quad t > 0,$$

where

$$I(x, t) = \sum_{j=1}^{n_i} I_{ij}(x, t), \quad i = 0, 1, \cdots, m, \quad x_i < x < x_{i+1}$$

and V satisfies the two-point boundary conditions

$$\left[\alpha V(x, t) + b \frac{\partial V(x, t)}{\partial x} \right]_{x=0} = c(t), \quad t > 0,$$

$$\left[\alpha V(x, t) + \beta \frac{\partial V(x, t)}{\partial x} \right]_{x=l} = \gamma(t), \quad t > 0,$$

with initial condition

$$V(x, 0) = v(x), \quad 0 < x < l,$$

where

$$\gamma(t) = \sum_{j=1}^{n} \left(\frac{d_{mj}}{d_{01}} \right)^{3/2} \gamma_j(t),$$

and

$$v(x) = \sum_{j=1}^{n_i} \left(\frac{d_{ij}}{d_{01}} \right)^{3/2} v_{ij}(x), \quad i = 0, 1, \cdots, m, \quad x_i < x < x_{i+1}.$$

The proof of this result is quite straightforward. One just checks that V satisfies the cable equation (1.2) on each interval (x_i, x_{i+1}). The constraints of continuity of electrical potential and conservation of current along with the three-halves power law (1.1) guarantee that $\partial V/\partial x$ is continuous at branch points. The proof is finished by taking Laplace transforms which convert the coupled partial differential equations to ordinary ones.

The mapping Theorem 1.1 is useful because, instead of having to solve several cable equations and match their solutions at branch points, one has to solve only one such equation on $(0, l)$. This is often easy using Green's functions. Also, on $(0, x_1)$ the mapping is one-to-one so that the potential on the trunk may be found immediately. One may also treat the case of several dendritic trees joined at a soma. For examples and further discussion, see Walsh and Tuckwell (1985).

If one wishes to use the cable equation to model the electrical potential of a nerve cell and to include the possibility of action potentials, then as with the Lapicque model, one must arbitrarily choose a threshold condition. Some are discussed in Chapter 6.

Nonlinear reaction-diffusion systems. From their experimental results on ionic currents for squid axon under voltage clamp, Hodgkin and Huxley (1952) formulated a system of nonlinear reaction-diffusion equations with the four components V, n, m, and h, which are the voltage, potassium activation, sodium activation, and sodium inactivation, respectively. Only one of the four equations had diffusion in it:

$$\frac{\partial V}{\partial t} = \frac{\partial^2 V}{\partial x^2} + g_K(V_K - V) + g_{Na}(V_{Na} - V) + g_l(V_l - V).$$

Here g_K, g_{Na}, and g_l are the potassium conductance, sodium conductance, and leakage conductance; V_K, V_{Na}, and V_l are the corresponding Nernst potentials. These are discussed further in Chapter 6. Unlike the Lapicque and cable models, this system has *natural threshold* properties in the sense that a strong enough excitatory stimulus will lead to the formation and propagation of an action potential.

Because the Hodgkin–Huxley equations are complicated, simplified nonlinear reaction-diffusion systems with similar properties have been employed for the purpose of analysis. One of these is the two-component system called the Fitzhugh–Nagumo equations, in which there is a voltage variable and a recovery variable (see Chapter 6). A nonlinear system of ordinary differential equations has been found superior to the Hodgkin–Huxley equations for nodal membrane. These are the Frankenhaeuser–Huxley (1964) equations. Both the Fitzhugh–Nagumo and Frankenhaeuser–Huxley equations also have natural threshold properties.

1.4. Stochastic phenomena in neurobiology.

We briefly summarize experimental studies of neurons and nervous systems in which the observed phenomena of interest are stochastic. These are given in approximately the chronological order of their discoveries.

1.4.1. The electroencephalograph (EEG). In 1875, Caton made the first recordings of electrical activity from skulls of animals. The observed small voltage fluctuations are known as the electroencephalograph (EEG). Similar recordings may be made directly from a brain surface. The first human EEG was obtained in 1924 and published five years later (Berger (1929)). EEGs are used often in clinical studies. Some analyses of the EEG as a random process are discussed in Chapter 10.

1.4.2. Evoked potentials (EP). Evoked potentials were first studied in 1913 by Pravdich–Neminsky, who showed that an electrical shock delivered to the sciatic nerve of a dog gave rise to a transient change in the EEG. Evoked potentials may be elicited by stimuli in many sense modalities and are also used routinely in clinical tests for pathological nervous conditions. For example, they are useful in diagnosing multiple sclerosis. Like the EEG, EPs involve populations of nerve cells and are discussed in Chapter 10.

1.4.3. Fluctuations in excitability. In 1932, Blair and Erlanger reported that identical electric shocks elicited action potentials in axons in a random fashion. This was interpreted as being due to fluctuations in threshold; quantitative results were given by Pecher (1939).

1.4.4. Variability in the interspike interval (ISI). In 1946, Brink, Bronk, and Larrabee recorded sequences of action potentials from frog muscle spindles and found that the time intervals between them were random. This was confirmed later in many studies. One of the first experiments that revealed how variable the ISIs were in CNS cells was that of Frank and Fuortes (1955) on cat spinal neurons. A significant advance in the quantitative analysis of ISI distributions was made by Gerstein and Kiang (1960). Since then, ISI randomness, not only in spontaneous activity (if it occurs) but also in driven activity, has been found in every cell examined. Models for this variability are discussed in Chapters 3–6. Studies of ISI variability are also an important component of theories of learning and perception. Weak signals must be distinguished from background noise. Correlation studies are useful in the analysis of multi-unit activity, and recordings may be made simultaneously from 20 or more cells. Statistical analysis is presented in Chapter 7.

1.4.5. Miniature end-plate potentials. Fatt and Katz (1950) made a detailed study of spontaneously occurring small synaptic potentials at neuromuscular junctions. The times of occurrence were postulated to conform to a Poisson point process. The quantum hypothesis was made that end-plate potentials were a random number of the miniature potentials (see Chapter 2).

1.4.6. $1/f$ noise. Verveen and Derksen (1965) studied voltage fluctuations near resting level in frog nodal membrane. The noise spectrum was inversely proportional to frequency at high frequencies. This phenomenon is found in semiconductors and is called $1/f$ ("one over f") noise.

1.4.7. Channel noise. Voltage and current fluctuations in membranes due to the random opening and closing of ionic channels were observed and analyzed by Katz and Miledi (1970), who saw how such studies could lead to estimates of microscopic quantities. Investigations of this so-called channel noise received a large impetus from the advent of single-channel recordings (Neher and Sakmann (1976); see Chapter 8).

1.4.8. Random stimulation. Neurophysiologists often employ random inputs to obtain nerve cell properties. One method is called *system identification*; in it a network is characterized by its response to white noise. This leads to Wiener kernel expansions (see Chapter 9).

General references for, and reviews of, the present and related subject matter are Burns (1968); Holden (1976); Jack, Noble, and Tsien (1985); and Tuckwell (1988a,b).

CHAPTER 2

Synaptic Transmission

We will briefly consider two of the manifestly stochastic features of the transmission of signals from nerve to nerve or from nerve to muscle. One of these is the *amplitude* of the response in the postsynaptic cell, and the other concerns the *timing* of the spontaneous postsynaptic potentials. The underlying physiology and morphology for the process of synaptic transmission has not been firmly established despite over 30 years of intensive labor.

2.1. Amplitude.

In 1936, Dale and his colleagues found convincing evidence that the transmission of signals from nerve to a certain muscle was mediated by the chemical acetylcholine (Dale, Feldberg, and Vogt (1936)). The mechanisms by which this transmission occurs, however, were not known. A clue was provided by the Fatt and Katz (1950) discovery of a spontaneous transmission at frog neuromuscular junction with a rate of about one per second. The amplitudes of these spontaneous potential changes were much smaller than those resulting from the response evoked by a nerve impulse. The latter is called an end-plate potential (EPP); the smaller spontaneous responses are called miniature EPPs, or simply MEPPs.

The MEPP amplitudes were found to be random with a mean of about 0.5 mV. The *quantum hypothesis* of del Castillo and Katz (1955) claimed that an EPP was an integral number of MEPPs. That is, synaptic transmission occurred in multiples of some unit or quantum. At about the same time, the electron microscope (EM) was emerging as a powerful anatomical tool, and EM studies revealed the existence of spheroidal vesicles of approximate diameter 500 Å in the presynaptic terminals. These were found to contain acetylcholine. The *vesicle hypothesis* identified the quantum of synaptic transmission with the release of one vesicle into the synaptic cleft.

A simple *stochastic model* was advanced which could test the quantum hypothesis. Based on the anatomical evidence, it was supposed that there were physically distinguishable transmitter release sites. When a nerve impulse invades the junction, a random number (say, N) of these sites are activated. Let the contributions from the various release sites be independent and identically distributed random variables $\{X_k\}$ so that the total response is

$$V = X_1 + X_2 + \cdots + X_N.$$

Based on experimentation, a reasonable approximation is that the X_k's are normally distributed with mean μ and variance σ^2, although gamma densities are perhaps a better choice (Robinson (1976)). The natural choice for N is that of a binomial random variable, but in the original model it was assumed to be Poisson (λ), so the amplitude of the EPP becomes *compound Poisson*. The density of V is then given by

$$f(v) = e^{-\lambda}\left[\delta(v) + \frac{1}{\sqrt{2\pi\sigma^2}}\sum_{k=1}^{\infty}\frac{\lambda^k}{k!\sqrt{k}}\exp\left\{\frac{-(v-k\mu)^2}{2k\sigma^2}\right\}\right],$$

where $\lambda = E(N)$ (this formula was first given by Bennett and Florin (1974)). The mean and variance of V are

$$E(V) = \lambda\mu, \quad \text{Var}(V) = \lambda(\mu^2 + \sigma^2).$$

The excellent fit of $f(v)$ to experimental distributions of some EPP amplitudes provided impressive evidence to support the quantum hypothesis (see, e.g., Boyd and Martin (1956)). The details of the release processes, however, remain controversial (Trembloy, Laurie, and Colonnier (1983)). Several authors have addressed the problem of parameter estimation (Miyamoto (1975); Brown, Perkel, and Feldman (1976); Robinson (1976)). It must be emphasized that what is found at neuromuscular junctions may not apply to other synapses.

2.2. Timing.

Fatt and Katz (1952) collected the irregularly spaced time intervals between MEPPs into a histogram and applied a χ^2 goodness-of-fit test for an exponential distribution. The result led to the *Poisson hypothesis*, namely that the occurrence times of MEPPs were generated by Poisson process with constant intensity.

This simple hypothesis is not without ramifications. If it is true, it points to the independence of the releases at a given site and from various sites. Thus departures from the Poisson hypothesis would have to be explained by some physiological mechanism. For example, the drag hypothesis, that one release tends to promote one nearby in space or time, would lead to a clustering of recorded MEPP times.

To construct a realistic model of the release process, including the anatomical details of the synaptic junction, seems very difficult. Simple queueing models such as those of Vere-Jones (1966) are nevertheless not without usefulness.

Some physiologists have taken the Poisson hypothesis as a possible absolute law of nature. This has led to studies in which it has been confirmed and rejected under various experimental conditions and with various statistical tests (see, e.g., Van der Kloot, Kita, and Cohen (1975)). One benefit has been progress with tests for Poisson processes, commencing with the work of Lewis (1965) and Cox and Lewis (1966).

More recently, the original hypothesis has been extended to include Poisson processes with time-dependent intensity:

$$\lambda(t) = \lim_{\Delta t \to 0} \frac{1}{\Delta t} \Pr\{\text{an event in } (t, t + dt]\}, \qquad t > 0.$$

The analysis of such processes is facilitated by the fact that they may be transformed to a standard temporally homogeneous Poisson process by the change of time scale,

$$\tau = \int_0^t \lambda(s)\, ds.$$

Maximum likelihood methods may be employed to estimate $\{\lambda(t)\}$ (Cox and Hinkley (1975)). Details may be found in Yana et al. (1984).

There are other stochastic aspects of synaptic transmission. One of these concerns *channel noise* at postsynaptic membrane, but this is dealt with in Chapter 8. Another concentrates on the temporal relationships between presynaptic spike trains and the spike trains in a postsynaptic cell. This methodology, called *synaptic "identification,"* was introduced by Brillinger (1975) and Krausz (1975) and is similar in principle to Wiener's expansion, which is discussed in Chapter 9.

CHAPTER 3

Early Stochastic Models for Neuronal Activity

In this chapter, we will consider some of the historical development of stochastic neuronal modeling, beginning with Landahl (1941). Simple models, such as Poisson processes and random walks, will be briefly explored, including the mathematical justification for treating the train of input spikes to a spontaneously active cell as a Poisson process. For the latter we rely on a paper by Cinlar (1972).

It should be kept in mind that the first recordings of EPSPs and IPSPs did not occur until 1952, so that models invented before that time were inclined to concentrate purely on the spatial summation of inputs—although it was known that temporal summation also existed. The McCulloch–Pitts (deterministic) model, with its discretization of time and (in hindsight) rather peculiar summation rules, until quite recently paved the way for work on stochastic neuronal models, which seem to have little relevance to the manner in which real neurons integrate their inputs.

3.1. Early models.

Landahl (1941) was perhaps the first to consider a stochastic model of neural response. One of his aims was to obtain a quantitative theory for the results of Pecher (1939) which we mentioned in Chapter 1. A condition for firing in Rashevsky's (1938) two-factor nerve model was stochasticized by replacing the net excitation by a normal random variable with mean equal to the deterministic value.

With a few additional assumptions, Pecher's experimental results were reasonably fitted, but it is now known that the underlying model is inappropriate and that the randomization procedure is too ad hoc to be acceptable. Interestingly, however, Landahl did broach the problem of determining the distribution of firing times in the case of a constant input current.

In the same year that McCulloch and Pitts (1943) wrote "A logical calculus of the ideas immanent in nervous activity," there appeared a stochastic model for neuron firing (Landahl, McCulloch, and Pitts (1943)). In this model, a neuron fires if at least n_θ excitatory inputs occurred within a time interval of length δ and there were no inhibitory inputs in that time interval. There were assumed to be $n_p \geq n_\theta$ excitatory synapses on the given cell together with a certain number of inhibitory inputs, all of which were governed by Poisson processes. Similar comments apply to this model as to that of Landahl.

McCulloch and Pitts (1948) were concerned with the fact that the nervous system had to deal with continua even though neurons appeared to be "all or none" in their signaling. They devised a model for spinal motoneurons receiving excitation from stretch receptors, as follows.

Let the average frequency of firing of each receptor be ρ and interpret $\rho \Delta t$ as the probability that a receptor fires in a time interval of length Δt. Henceforth take $\Delta t = 1$ and assume $0 < \rho \leq 1$. Suppose that n receptors make synaptic contact with a motoneuron, that receptors act independently, and that n is large enough to apply the central limit theorem. Then the number of inputs received by a motoneuron in a unit time interval is

$$X \stackrel{d}{=} N(n\rho, \sqrt{n\rho(1-\rho)}),$$

where $N(\mu, \sigma)$ signifies a normal random variable with mean μ and standard deviation σ. The threshold condition is that a cell fires if it receives h or more inputs in a unit time interval. Hence

$$p(\rho) = \Pr\{\text{motoneuron fires in a unit time interval}\}$$
$$= \Pr\{X \geq h\}$$
$$\simeq \frac{1}{\sqrt{2\pi}} \int_{(h-n\rho)/\sqrt{n\rho(1-\rho)}}^{\infty} \exp(-x^2/2)\, dx.$$

Since $p(\rho)$ gives the average output frequency, the latter is, assuming $n > h$, a sigmoid function of ρ with $p(0) = 0$ and $p(1) = 1$. It is noteworthy that McCulloch and Pitts ended this paper with, "··· [T]he mathematical treatment of the activity of the nervous system presents numerous problems in the theory of probability and stochastic processes."

Rosenbleuth et al. (1949) were also concerned with the input-output relations of spinal motoneurons. It was assumed that dorsal root (afferent) fibers synapse randomly with motoneurons whose axons leave the spinal cord in the ventral root (efferent fibers). If a stimulus is delivered to the dorsal root, a proportion of the afferent fibers are excited. Let there be n afferent fibers synapsing on each motoneuron. Then the probability of k active inputs to a given cell is $\binom{n}{k} p^k (1-p)^{n-k}$ and, on the further assumption that there must be at least m simultaneous excitations for a motoneuron to fire, we have

$$p_m = \Pr\{\text{a given motoneuron fires}\}$$
$$= \sum_{k=m}^{n} \binom{n}{k} p^k (1-p)^{n-k}.$$

Thus in a population of N motoneurons

$$E(\text{number of motoneurons firing}) = \sum_{j=0}^{N} j \binom{N}{j} p_m^j (1-p_m)^{N-j}.$$

Using a normal approximation when it should have been accurate led to disagreement with the experimental observations on ventral root discharges as a function of the strength of the dorsal root volleys. This forced Rosenbleuth et al. (1949) to modify the above model by including variability in thresholds for firing in the motoneuron pool. Interestingly, the value of n was set at about 29, which can be compared with the estimated figure of 32 from more recent data (Koziol and Tuckwell (1978)). The statistical aspects of monosynaptic input-output relations for motoneurons were also considered by Rall (1955a,b).

3.2. Neural firing as a first passage time.

In this section we give a general scheme of a starting point for simple stochastic neural models. In Chapter 1 we saw that the depolarization $V(x, t)$ over a neuron cylinder may be given approximately, in appropriate units, by solutions of the cable equation

$$V_t = V_{xx} - V + I, \quad a < x < b, \quad t > 0,$$

and with given boundary conditions. Taking a spatial average of the potential, defined by

$$\bar{V}(t) = \frac{1}{b-a} \int_a^b V(x, t) \, dx,$$

we find that $\bar{V}$ satisfies

$$\frac{d\bar{V}}{dt} = \frac{1}{b-a}(V_x(b, t) - V_x(a, t)) - \bar{V} + \bar{I}, \quad t > 0,$$

where $\bar{I}$ is the average current density. Assume sealed end conditions so that there is no escape of longitudinal current at the ends of the cylinder. Then $V_x(a, t) = V_x(b, t) = 0$, and we are left with the equation for the potential V (dropping the overbar) in the Lapicque (1907) model:

(3.1) $$\frac{dV}{dt} + V = I.$$

This equation is the starting point for some models considered in Chapters 4 and 5. Our immediate concern is to simplify matters even further. If we recast (3.1) in terms of an R-C circuit with time constant τ, then it is written

$$\frac{dV}{dt} + \frac{V}{\tau} = \frac{I}{C}.$$

Here any initial potential $V(0)$ decays exponentially as $V(0)e^{-t/\tau}$. A further simplification results by assuming $\tau = \infty$ so that there is no decay and, in fact, in the absence of any threshold effects,

(3.2) $$V(t) = V(0) + \frac{1}{C} \int_0^t I(t') \, dt'.$$

This model, devoid of any physiological or anatomical realities, is called the *perfect integrator*. There is perfect summation of all inputs. Although it might seem an extremely crude approach, the use of the perfect integrator as a starting point for stochastic neuronal models is useful because it introduces simple processes for which exact results are often obtainable. Then we may slowly introduce more details with concomitant increases in mathematical complexity.

3.2.1. Random input. Let us assume that the input current I is a random process; then $\{V(t)\}$ is a random process related to $\{I(t)\}$ by (3.2). This statement requires modification if the action potentials are to occur. The simplest *threshold assumption* is that when $V(t)$ exceeds a threshold θ, which may be time varying or constant, the neuron emits an action potential. This will usually be followed by an *absolute refractory period* of duration t_R, during which the cell is incapable of emitting another action potential. In one-dimensional models it is usually supposed that incident excitation or inhibition is ineffective during this period, which is approximately valid in most cases.

The first *interspike interval* is then the *first passage time* random variable

$$T_1 = \inf\{t : V(t) \geq \theta(t)\},$$

the value ∞ being assigned to T_1 if the indicated set is empty. At time $t_R + T_1$ the potential is reset, usually to $V(0)$, which is often taken as the resting potential. (A possible refinement is to reset V to a less excited value to take account of the phenomenon of *afterhyperpolarization*, but this is an ad hoc procedure.) Then we wait for a second threshold crossing after an interval of length T_2, which means that the second interspike interval is $t_R + T_2$. The random variables $T_1, T_2, \cdots$ are in the first approximation and assumed to be independent and identically distributed so that the train of spikes is a *renewal process*.

These are the ideas behind many one-dimensional models, which differ only in their assumptions about the nature of the process $\{V(t)\}$, and the threshold function $\theta(t)$.

3.3. Poisson processes as approximations to input event sequences.

The membrane potential at a point on a neuron's surface is continuous in time, even during an action potential. The most significant part of the latter is an abrupt depolarization, which is often so brief compared to the time interval between action potentials that we may extract from the train of spikes a sequence of time points (for example, the set of times at which the potential crossed a given level in the positive direction). Thus a random spike train can be viewed as a realization of a *point process*.

Ignoring differences in the spatial origins of the various synaptic inputs to a given cell, we may suppose that there are n_E excitatory input channels and n_I inhibitory input channels. If the times at which synaptic inputs arrive on any of these $n = n_E + n_I$ channels are considered to be point processes, then the total input sequence is a *superposition* of n point processes. A neuron may then be viewed as operating on a random point process to produce another, usually in a highly nonlinear fashion.

As a starting point in the theory of neuronal integration of random inputs, it is convenient to assume that the pooled excitatory inputs constitute a Poisson process N_E and that the pooled inhibitory inputs constitute another Poisson process N_I. There is substantial mathematical justification for this, as we see in the following results. The general theme of these results is that a superposition of a large number of uniformly sparse point processes is approximately a Poisson process. More specific results are possible when the individual processes are renewal processes.

Weak convergence in the next theorem means convergence of finite-dimensional distributions. Also, a counting random measure N on $(R^+, \mathcal{B}(R^+))$ is Poisson with mean measure μ if, for disjoint $A_1, \cdots, A_m \in \mathcal{B}(R^+)$, the random variables $N(A_1), \cdots, N(A_m)$, representing the numbers of points in the respective subsets, are independent and

$$\Pr\{N(A) = k\} = e^{-\mu(A)} \mu(A)^k / k!, \quad k = 0, 1, \cdots,$$

where $A \in \mathcal{B}(R^+)$ and $\mu(A) < \infty$. We can now state the main result.

THEOREM 3.1. *Let*

$$M_{11}, \cdots, M_{1k_1},$$
$$M_{21}, \cdots, M_{2k_2},$$
$$\vdots$$
$$M_{n1}, \cdots, M_{ni}, \cdots, M_{nk_n},$$

where $k_n \to \infty$ as $n \to \infty$, be a double sequence of point processes such that
(i) *the processes in any row are mutually independent;*
(ii) *for any bounded interval B,*

$$\lim_{n \to \infty} \sup_{1 \le i \le k_n} \Pr\{M_{ni}(B) \ge 1\} = 0.$$

Then the superposition processes

$$M_n = M_{n1} + \cdots + M_{nk_n}$$

converge weakly as $n \to \infty$ to a Poisson process with mean measure μ if and only if

$$\lim_{n \to \infty} \sum_{i=1}^{k_n} \Pr\{M_{ni}(B) = 1\} = \mu(B)$$

and

$$\lim_{n \to \infty} \sum_{i=1}^{k_n} \Pr\{M_{ni}(B) \ge 2\} = 0$$

for any bounded interval $B \in \mathcal{B}(R^+)$.

The above theorem was given by Franken (1963) and Grigelionis (1963); it is also proved in Cinlar (1972). Since we may often wish to consider spike trains that are stationary, the following corollary is also useful. For this we assume that each of the component processes in the above theorem is stationary, the *intensity* of M_{ni} being denoted by λ_{ni}. Furthermore, it is assumed that

$$\lim_{n\to\infty} \sup_{1\le i\le k_n} \lambda_{ni} = 0.$$

COROLLARY 3.2. *The sequence of processes* $\{M_n\}$ *converges weakly to a stationary Poisson process with intensity* λ *if and only if*

$$\lim_{n\to\infty} \sum_{i=1}^{k_n} \lambda_{ni} = \lambda$$

and

$$\lim_{n\to\infty} \sum_{i=1}^{k_n} \Pr\{M_{ni}(t) \ge 2\} = 0, \qquad t > 0,$$

where $M_{ni}(t)$ is the number of points of M_{ni} in $(0, t]$.

This corollary is also proved in Cinlar (1972).

Since many nerve cells have thousands of synapses, if a large number of these are active, we can appeal to Theorem 3.1 to justify the use of Poisson processes to approximate the arrival times of synaptic inputs, with separate considerations for excitatory and inhibitory inputs. In fact, we may be justified in using a Poisson approximation for the inputs of a given type—that is, whose presynaptic fibers have a common anatomical origin. However, there is not in general sufficient information on the origins of the synaptic inputs of the majority of CNS cells to make these claims. Clearly, the Poisson assumption is not valid when a cell is subjected to a highly regular stimulus, although there may be a Poisson scatter about a deterministic input. Furthermore, the input connections are fixed in number, if one ignores developmental and aging aspects, so that a limiting operation is not possible. One may then be interested in how accurate the Poisson approximation is, as this matter is also discussed in Cinlar (1972). However, the lack of information on the complex inputs to most CNS cells makes a detailed study seem unnecessary at present. It is worth mentioning that the density of the time interval between input events has been found to be decidedly not exponential in some cases, as, for example, in cat lateral geniculate neurons in spontaneous activity (Bishop, Levick, and Williams (1964)).

3.4. Neurons driven by Poisson excitation.

Let $\{N(t), t \ge 0\}$ be a simple Poisson process with constant intensity λ and $N(0) = 0$ almost surely. We suppose that a neuron receives an excitatory input each time an event occurs in $\{N(t)\}$. In the perfect integrator picture, we imagine that each excitatory input instantaneously delivers a certain charge to the neuron, so $I(t)$ consists of a sequence of delta functions. Let the depolarization jump by α at each input so that in the absence of a threshold

$$V(t) = \alpha N(t).$$

As is well known, $\lceil \theta/\alpha \rceil$ events are needed to reach θ, where $\lceil z \rceil$ is the ceiling of z, i.e., the smallest integer greater than or equal to z. Thus the firing time, T, including an absolute refractory period, t_R, has density

(3.3) $$f_T(t) = \begin{cases} 0, & 0 < t < t_R, \\ \dfrac{\lambda(\lambda(t-t_R))^{n-1}}{(n-1)!} e^{-\lambda(t-t_R)}, & t > t_R, \end{cases}$$

where $n = \lceil \theta/\alpha \rceil$.

The case $n = 1$ gives an exponential density. Any excitatory event triggers an action potential in the postsynaptic cell if the latter is nonrefractory. This does occur, for example, in certain cells of the cat cochlear nucleus (Gerstein (1962)) where there is good evidence to support the claim of a one-to-one correspondence between input and output (cf. Molnar and Pfeiffer (1968)). Exponentially distributed interspike times are also found in some neurons of the cat superior olivary complex under certain conditions of auditory stimulation (Goldberg, Adrian, and Smith (1964)). Although the densities for $n \geq 2$ in (3.3) resemble those for the ISIs of many neurons, based on known physiological and anatomical facts it is unlikely that they arise via the simple waiting time description we have given here.

3.4.1. Poisson excitation with linear decay. We have seen in Chapter 1 that the declining phase of EPSPs often consists of exponential-like decay. Since for small Δt we have

$$V(0)e^{-\tilde{\beta}\Delta t} \simeq V(0)[1 - \tilde{\beta}\Delta t],$$

it would seem reasonable to assume that in the case of a high frequency Poisson input the exponential decay might be regarded as approximately linear. In this model we assume:

(a) Poisson excitation at constant intensity λ, each jump in potential being of magnitude $\alpha > 0$;
(b) linear decay at constant rate $\tilde{\beta} \geq 0$ in the absence of inputs;
(c) a constant threshold $\tilde{\theta}$;
(d) the nerve cell is initially at rest, $V(0) = 0$.

Then for subthreshold levels of excitation,

$$V(t) = \alpha N(t) - \tilde{\beta} t$$

and the firing time is

$$T = t_R + T,$$

where

$$T = \inf\{t : V(t) \geq \tilde{\theta}\}$$
$$= \inf\{t : N(t) - \beta t \geq \theta\},$$

where

$$\beta = \tilde{\beta}/\alpha, \qquad \theta = \tilde{\theta}/\alpha.$$

The following result is due to Pyke (1959).

THEOREM 3.3.

$$\Pr\left\{\sup_{0\le t\le \tau} (N(t) - \beta t) \le \theta\right\} = e^{-\lambda\tau} \sum_{n=0}^{[\beta\tau+\theta]} \frac{(\beta\tau + \theta - n)}{n!} \left(\frac{\lambda}{\beta}\right)^n$$

$$\times \sum_{j=0}^{[\theta]} \binom{n}{j}(j-\theta)^j(\beta\tau + \theta - j)^{n-j-1},$$

where $\binom{n}{j} \doteq 0$ for $j > n$ and $[z]$ is the greatest integer less than z.

The result, of course, gives the distribution of firing times since

$$\Pr\{T \le \tau\} = 1 - \Pr\{T > \tau\}$$

$$= 1 - \Pr\left\{\sup_{0\le t\le \tau} (N(t) - \beta t) \le \theta\right\}.$$

Furthermore, we have the following corollary on the probability of firing in a finite time.

COROLLARY 3.4.

$$\Pr\{T < \infty\} = \begin{cases} 1 - \left(1 - \frac{\lambda}{\beta}\right) \sum_{j=0}^{[\theta]} \left(\frac{\lambda}{\beta}\right)^j \frac{(j-\theta)^j}{j!} e^{-\lambda(j-\theta)/\beta}, & \lambda < \beta, \\ 1 & \text{otherwise.} \end{cases}$$

These results are a useful approximation for the discontinuous model with exponential decay which is considered in the next chapter. Keilson (1963) obtained a first passage time result for a similar kind of process; unfortunately it does not apply to the neuronal model under consideration.

3.5. Random walks.

Continuing with the idea of a perfect integrator, we will consider some random walk models of neural activity. The first one is in discrete time.

3.5.1. Periodic random excitation. We suppose that the input is periodic and that successive EPSPs are independent and identically distributed. The following result may be proved graphically.

THEOREM 3.5. Let $\{X_k, k = 1, 2, \cdots\}$ be independent and identically distributed with common distribution function $F(x) = 1 - e^{-\lambda x}$, $x \ge 0$. Let

$$V(n) = \sum_{k=1}^{n} X_k, \quad n = 1, 2, \cdots,$$

with $V(0) = 0$. Then the time of first passage of V to θ

$$T = \inf\{n : V(n) \ge \theta\}$$

has the probability law

$$\Pr\{T = n\} = \frac{e^{-\lambda\theta}(\lambda\theta)^{n-1}}{(n-1)!}, \quad n = 1, 2, \cdots.$$

Exact results may be obtained for other than exponentially distributed jumps, but the formulas tend to be unwieldy. Of course, these are instances of the classic problem of first passage times for partial sums of independent and identically distributed random variables.

3.5.2. Randomized random walk.
The inclusion of inhibition makes the determination of first passage times more difficult. However, when inhibition and excitation have equal but opposite effects on the state of excitation, exact results are available in the perfect integrator framework. In the neural context, the difference between two Poisson processes was first employed by Gerstein (1962). Further developments were introduced by Gerstein and Mandelbrot (1964).

With time now continuous, the subthreshold depolarization is assumed to be given by

$$(3.4) \qquad V(t) = N_E(t) - N_I(t), \qquad t > 0,$$

where N_E and N_I are independent simple (unit jumps) Poisson processes with constant intensities λ_E and λ_I, respectively, and with $N_E(0) = N_I(0) = 0$, almost surely. The process $\{V(t)\}$ was called a "randomized random walk" by Feller (1966), who gave many of the results that are relevant in the present setting. The results are also derived in Tuckwell (1988b). It is assumed that the cell fires when V reaches a constant threshold θ for the first time.

In the *absence of barriers*, it is probable that the random walk is a positive integer at m, and at time t is

$$p_m(t) = \left(\frac{\lambda_E}{\lambda_I}\right)^{m/2} e^{-\lambda t} I_m(2t\sqrt{\lambda_E \lambda_I}),$$

where

$$\lambda = \lambda_E + \lambda_I,$$

and where $I_m(\cdot)$ is a modified Bessel function defined by

$$I_m(x) = \sum_{k=0}^{\infty} \frac{1}{k!\,\Gamma(k+m+1)} \left(\frac{x}{2}\right)^{2k+m}$$

To obtain f, the density of the time of first passage to $\theta > 0$,

$$T = \inf\{t : V(t) \geq \theta\},$$

we may utilize the *renewal equation*

$$p_m(t) = \int_0^t f(t') p_{m-\theta}(t-t')\,dt',$$

where it is assumed that $0 < \theta < m$. Laplace transforming this equation yields the Laplace transform of f. Upon inversion one obtains

$$(3.5) \qquad f(t) = \theta\left(\frac{\lambda_E}{\lambda_I}\right)^{\theta/2} \frac{e^{-\lambda t}}{t} I_\theta(2t\sqrt{\lambda_E \lambda_I}), \qquad t > 0.$$

The special case $\lambda_E = \lambda_I$ can also be treated using the method of images.

Stability. Gerstein (1962) and Gerstein and Mandelbrot (1964) claimed that one reason why they were alerted to the use of the randomized random walk model for certain neurons was a stability property of the first passage density (3.5) in the symmetric case $\lambda_E = \lambda_I$. However, the stability property of the theoretical firing time did not match experimental ones. The property found for certain cells of the cochlear nucleus was as follows.

Define T_m as the time interval between every 2^mth spike so that T_m is the sum of 2^m interspike intervals. Put $T_0 = T =$ an ISI, and let the corresponding densities be $f(t)$ for T, and $f_m(t)$ for T_m. Then experimentally

$$f(t) = 2f_1(2t) = 4f_2(4t) \cdots$$

was approximately true for a certain class of cells. This is not satisfied by (3.5). See Holden (1976) for a discussion of the possible usefulness of stable laws in relation to information processing.

Moments. Define the nth moment of T, if it exists, by

$$\mu_n = \int_0^\infty t^n f(t)\, dt.$$

From the general theory of birth and death processes we find that

$$\mu_0 = \Pr\{T < \infty\} = \begin{cases} 1, & \lambda_E \geq \lambda_I, \\ \left(\dfrac{\lambda_E}{\lambda_I}\right)^\theta, & \lambda_E < \lambda_I. \end{cases}$$

Thus if the intensity of excitation is greater than or equal to the intensity of inhibition, in this model, the neuron fires in a finite time with probability one. Otherwise, there is a nonzero probability that the cell will never emit an action potential.

The integrals for $n = 1$ and $n = 2$ can be done with the aid of the following integral (see, e.g., Gradshteyn and Ryzhik (1965)):

$$\int_0^\infty e^{-\alpha x} I_\nu(\beta x)\, dx = \frac{\beta^\nu}{\sqrt{\alpha^2 - \beta^2}(\alpha + \sqrt{\alpha^2 - \beta^2})^\nu}.$$

When $\lambda_E > \lambda_I$, this yields,

$$E(T) = \frac{\theta}{\lambda_E - \lambda_I}, \quad \mathrm{Var}(T) = \frac{\theta(\lambda_E + \lambda_I)}{(\lambda_E - \lambda_I)^3}.$$

When $\lambda_E = \lambda_I$, $\mu_n = \infty$ for $n \geq 1$.

Tails. Using the asymptotic property of the modified Bessel function for large arguments (see, e.g., Abramowitz and Stegun (1965)), we find that when there is Poisson excitation and Poisson inhibition,

$$f(t) \underset{t\to\infty}{\sim} \frac{\theta}{2}\left(\frac{\lambda_E}{\lambda_I}\right)^{\theta/2} \frac{1}{\sqrt{\pi}(\lambda_E\lambda_I)^{1/2}} e^{-t(\sqrt{\lambda_E}-\sqrt{\lambda_I})^2}\left\{1 - \frac{4\theta^2 - 1}{16t\sqrt{\lambda_E\lambda_I}} + o(t^{-1})\right\},$$

whereas when there is excitation only the exact result,

$$f(t) = \frac{\lambda_E^\theta t^{\theta-1} e^{-\lambda_E t}}{(\theta - 1)!},$$

applies if θ is a positive integer. Thus the behavior of the first passage time density at large t is quite different in the cases with and without inhibition for this model.

3.5.3. General jump distributions. In the randomized random walk of the previous subsection, jumps occur at rate $\lambda = \lambda_E + \lambda_I$, and the jump amplitude has a density

$$\phi(y) = [\lambda_E \delta(y - 1) + \lambda_I \delta(y + 1)]/\lambda.$$

We may extend this to an arbitrary density $\phi(y)$, $y \in R$, keeping a total jump intensity λ and superimposing a linear drift βt.

This gives a temporally homogeneous process with independent increments with no diffusion component—in fact, it is a compound Poisson process with drift. Assuming the process starts at zero, let $p(x, t)$ denote its probability density at t. Then, as is well known (see, e.g., Cox and Miller (1965)), p satisfies the integrodifferential equation

$$\frac{\partial p}{\partial t} = -\beta \frac{\partial p}{\partial x} + \lambda \int_{-\infty}^{\infty} [p(x - y, t) - p(x, t)] \phi(y) \, dy.$$

The solution for an unrestricted process is easily shown by probabilistic arguments to be

$$p(x, t) = e^{-\lambda t} \sum_{k=0}^{\infty} \frac{(\lambda t)^k}{k!} \phi_k(x),$$

where

$$\phi_k(x) = \int \phi(x - y) \phi_{k-1}(y) \, dy, \quad k = 1, 2, \cdots,$$

$$\phi_0(x) = \delta(x - \beta t),$$

as given by Kryukov (1976), who used a Wald-type identity to investigate first passage times. However, the results obtained are of limited use in the present context. Finally, we note that Gusak and Koralyuk (1968) found an integral equation for the generating function of the first passage time of a compound Poisson process.

3.6. Some other approaches.

There have been other approaches to theories of stochastic neuronal activity. For example, Stein (1967) advocated that when synaptically induced conductance changes are long lasting, it is more convenient to consider current changes in the postsynaptic neuron rather than voltage changes. To conclude this chapter, we will consider briefly two other early models.

Hagiwara (1954) studied the variability of ISIs in fibers originating in frog muscle spindles. The source of randomness was assumed to be "thermal agitation," and at a given tension the state of the cell at time t after the previous action potential was given by

(3.6) $$S + X(t),$$

where S = constant, $X(t)$ = a "random normal process, the mean being zero and the standard deviation σ being constant." This is interpreted as meaning X is a mean zero stationary Gaussian process. A threshold function

$$\theta(t) = Ae^{c/t}, \quad c, A > 0, \quad t > 0$$

was employed with a built-in highly (but not absolutely) refractory state after a nerve firing. A first passage of $S + X(t)$ to $\theta(t)$ would result in a spike, followed by a *threshold reset*. This above choice of a threshold and a theoretical spike-generating condition are ad hoc procedures. A similar approach was adopted by Goldberg, Adrian, and Smith (1964) for certain cells of the cat superior olivary complex. The above threshold was modified to incorporate an absolutely refractory period, and $\{X(t)\}$ was assumed to be "Gaussian noise passed through an ideal low pass filter $\cdots$." By simulation, Goldberg, Adrian, and Smith were able to reproduce the statistical properties of the ISIs for the cells under consideration. An interesting observation was made: the same pattern of excitation could lead to quite different output patterns in different neurons. Further developments along these lines are given in Geisler and Goldberg (1966).

To illustrate both the variety of neuronal models and the fact that different cell types may require different approaches, we turn to pacemaker neurons in *Aplysia californica*. Such cells have ISIs with very small coefficients of variation (≈ 0.08). Junge and Moore (1966) examined intracellular voltage trajectories in these cells and found that the variability of the ISI was apparently due to varying voltage paths. The *threshold* was *fixed*. After each spike there was a reset of a single parameter, namely, the asymptotic voltage in the absence of threshold. Thus the kth voltage trajectory was

$$V_k(t) = C_k(1 - e^{-t/\tau}),$$

where $\{C_k\}$, $k = 1, 2, \cdots$ are independent and identically distributed and $\tau > 0$ is fixed. The kth ISI is obtained by solving $V_k(t) = \theta$ for t and adding an absolute refractory period.

Finally, we mention another approach taken for cat lateral geniculate nucleus (LGN) cells by Bishop, Levick, and Williams (1964). These neurons receive monosynaptic excitation from retinal ganglion cells. For one kind of cell it was assumed that an event in any input channel would lead to an LGN cell spike. The time intervals between spikes in individual input fibers were taken to be gamma variates. With n such independent input channels, the LGN cell should have a distribution function given by

$$F_n(t) = 1 - (1 - F(t))^n,$$

where F is common (gamma) distribution function of the input ISIs. A second class of LGN cells was assumed to be innervated by only one excitatory input fiber, as well as by a few inhibitory ones. The postsynaptic neuron would fire when an excitatory event occurred, but such events occasionally would be blocked by the activity in the inhibitory fibers. This should lead to the multimodal type ISI densities observed for this kind of cell.

In conclusion, we point out that we have by no means covered all the different kinds of early models. Some further discussion can be found in Holden (1976) and Lee (1979). Nor have we covered all the mathematical possibilities; for example, the optional stopping theorem for martingales can be used to quickly deduce mean first passage times for independent increment processes (see, e.g., Kannan (1979)).

CHAPTER 4

Discontinuous Markov Processes with Exponential Decay

In this chapter we will be primarily concerned with a model for stochastic neural activity proposed by Stein (1965). In this framework, the neuronal potential between action potentials is a discontinuous Markov process for which it has proven difficult to obtain analytical expressions for various quantities of interest. Because of these difficulties, it was natural that diffusion approximations be considered; these are the subject of Chapter 5.

4.1. Stein's model.

It was posited a long time ago that many neurons acted like R-C circuits (Lapicque (1907), Eccles (1957)). Indeed we found that if we averaged over the length of a nerve cylinder, then we obtained the equation $\dot{V} + V/\tau = I/C$. Observations of postsynaptic potentials made at a neuron's soma support this picture because they usually have short phases away from resting potential, followed by exponential-like decay back to resting potential. A typical time constant of decay is about 5–10 msec in the mammalian CNS. *Reversal potentials* complicate matters and are dealt with in §4.5.

Stein's model is a stochastic version of the R-C circuit model and thus does incorporate a realistic feature of nerve cell physiology, namely, the exponential decay just mentioned. Just as in the randomized random walk model, synaptic inputs occur as events in Poisson processes.

Let $V(t)$ be the depolarization at time $t \geq 0$, and let $N_E(t)$, $N_I(t)$, $t \geq 0$ be independent simple (unit jump) Poisson processes with intensities λ_E, λ_I, respectively, assumed to be constant. Let a_E, a_I be nonnegative real numbers. Then the basic model is as follows:
  (i) The cell is initially at rest so that $V(0) = 0$.
  (ii) Between jumps in N_E and N_I, V decays according to
$$\frac{dV}{dt} = -\alpha V,$$
where $\alpha = 1/\tau$ is the reciprocal of the membrane time constant.
  (iii) If a jump occurs in N_E at time t_0, then
$$V(t_0^+) - V(t_0^-) = a_E,$$

whereas an inhibitory input gives
$$V(t_0^+) - V(t_0^-) = -a_I.$$

(iv) When $V(t)$ attains a value θ (the threshold), the cell fires. In the first instance we assume that θ is constant.

(v) Upon firing, an absolute refractory period of duration t_R ensues. Thereupon V is reset to zero along with N_E and N_I, and the process starts again.

It can be seen that the sequence of output spikes thus constitutes a renewal process.

The subthreshold depolarization in the above model may be described by the equation
$$dV(t) = -\alpha V(t)\, dt + a_E\, dN_E(t) - a_I\, dN_I(t).$$

Before analyzing the solutions in relation to neuron firing, we will summarize the general theory of Markov processes, which will be applied to the above model in §4.3.

4.2. Markov processes as solutions of stochastic differential equations.

Stochastic differential equations for Markov processes with jump and diffusion components were first considered by Ito (1951b). An extensive report on the properties of vector Markov processes with deterministic, diffusive, and jump components is given in Gihman and Skorohod (1972). The theorems contained therein usually involve lengthy technical assumptions concerning measurability, etc., which we omit; we direct the interested reader to the last reference. It will be seen that such stochastic differential equations provide a succinct description of many mathematical models for nerve membrane potential under random influences.

A general stochastic differential takes the form
$$d\mathbf{X}(t) = \boldsymbol{\alpha}(\mathbf{X}(t), t)\, dt + \boldsymbol{\beta}(\mathbf{X}(t), t)\, d\mathbf{W}(t) + \int \boldsymbol{\gamma}(\mathbf{X}(t), t, \mathbf{u}) N(dt, d\mathbf{u}),$$

which is an abbreviation for the integral equation

(4.1)
$$\mathbf{X}(t) = \mathbf{X}(s) + \int_s^t \boldsymbol{\alpha}(\mathbf{X}(\tau), \tau)\, d\tau + \int_s^t \boldsymbol{\beta}(\mathbf{X}(\tau), \tau)\, d\mathbf{W}(\tau) \\ + \int_s^t \int \boldsymbol{\gamma}(\mathbf{X}(\tau), \tau, \mathbf{u}) N(d\tau, d\mathbf{u}), \qquad s \le t.$$

Here $\{\mathbf{X}(t)\}$ is a random process with values in R^n (or any n-dimensional Euclidean space); $\boldsymbol{\alpha}$ and $\boldsymbol{\gamma}$ are n-dimensional functions of their indicated arguments; $\boldsymbol{\beta}$ is an $n \times m$ matrix function; $\{\mathbf{W}(t)\}$ is an m-dimensional Wiener process whose individual components are independent standard Wiener processes, $\mathbf{u} \in R^n$, $t, s \in R^+$; and N is a Poisson random measure, independent of $\mathbf{W}$, defined on Borel sets of $R^+ \times R^n$ such that for $t_1 \le t_2$, $A \in \mathscr{B}(R^n)$, $N((t_1, t_2], A) \doteq N((t_1, t_2] \times A)$ is Poisson distributed with parameter
$$E[N((t_1, t_2], A)] = (t_2 - t_1)\Lambda(A).$$

We call Λ the rate measure and note that for fixed A, $\{N(t, A)\} \doteq \{N((o, t], A)\}$ is a Poisson process with intensity $\Lambda(A)$. For disjoint $A_1, \cdots, A_n$, $N(t, A_1), \cdots, N(t, A_n)$ are mutually independent. The definitions of stochastic integrals with respect to Wiener processes and Poisson random measure can be found in Gihman and Skorohod (1972). We point out that the integrals involving $\mathbf{W}$ are in accordance with Ito's definition.

The *infinitesimal generator* $\mathscr{A}_s$ of $\{\mathbf{X}(t)\}$, an integrodifferential operator acting on suitable scalar functions $f(\mathbf{x})$, $\mathbf{x} \in R^n$, is defined as follows. Let $\mathbf{X}(s) = \mathbf{x}$. Then

$$(\mathscr{A}_s f)(\mathbf{x}) = \lim_{\Delta s \to 0} \frac{E[f(\mathbf{X}(s + \Delta s)) - f(\mathbf{X}(s))]}{\Delta s}.$$

It can be shown that the infinitesimal generator of (4.1) is given by

(4.2)
$$(\mathscr{A}_s f)(\mathbf{x}) = \sum_{k=1}^{n} \alpha_k(\mathbf{x}, s) \frac{\partial f(\mathbf{x})}{\partial x_k} + \frac{1}{2} \sum_{k,l=1}^{n} (\boldsymbol{\beta}(\mathbf{x}, s) \boldsymbol{\beta}^T(\mathbf{x}, s))_{kl} \frac{\partial^2 f(\mathbf{x})}{\partial x_k \partial x_l}$$
$$+ \int [f(\mathbf{x} + \boldsymbol{\gamma}(\mathbf{x}, s, \mathbf{u})) - f(\mathbf{x})] \Lambda(d\mathbf{u}),$$

where $\boldsymbol{\beta}^T$ is the transpose of $\boldsymbol{\beta}$ so that

$$(\boldsymbol{\beta}(\mathbf{x}, s) \boldsymbol{\beta}^T(\mathbf{x}, s))_{kl} = \sum_{p=1}^{m} \beta_{kp}(\mathbf{x}, s) \beta_{lp}(\mathbf{x}, s).$$

The transition probability function of $\{\mathbf{X}(t)\}$ is

$$P(A, t; \mathbf{x}, s) = \Pr\{\mathbf{X}(t) \in A \mid \mathbf{X}(s) = \mathbf{x}\}$$

and for fixed A and t satisfies a *backward Kolmogorov equation*

(4.3)
$$\frac{\partial P}{\partial s} + (\mathscr{A}_s P(A, t \mid ., s))(\mathbf{x}) = 0.$$

If it exists, the transition probability density function p, defined through

$$P(A, t \mid \mathbf{x}, s) = \int_A p(\mathbf{y}, t \mid \mathbf{x}, s) \, d\mathbf{y},$$

under certain assumptions also satisfies (4.3). Furthermore, $p(\mathbf{y}, t \mid \mathbf{x}, s)$ satisfies a forward equation

(4.4)
$$\frac{\partial p}{\partial t} + \sum_{k=1}^{n} \frac{\partial}{\partial y_k} (\alpha_k(\mathbf{y}, t) p) - \frac{1}{2} \sum_{k,l=1}^{n} \frac{\partial^2}{\partial y_k \partial y_l} [(\boldsymbol{\beta}(\mathbf{y}, s) \boldsymbol{\beta}^T(\mathbf{y}, s))_{kl} p]$$
$$+ \lambda p - \int p(\mathbf{y} - \boldsymbol{\gamma}^*(\mathbf{y}, t, \mathbf{u}), t \mid \mathbf{x}, s) J(\mathbf{y}) \Lambda(d\mathbf{u}) = 0,$$

where

$$\lambda = \int \Lambda(d\mathbf{u}) < \infty$$

is the total jump rate. Here, if

$$\mathbf{z} = \mathbf{y} + \gamma(\mathbf{y}, s, \mathbf{u}),$$

then γ^* is defined through the inverse relation

$$\mathbf{y} = \mathbf{z} - \gamma^*(\mathbf{z}, s, \mathbf{u}),$$

and $J(\mathbf{z})$ is the Jacobian of this transformation. Proofs of (4.3) and (4.4), along with existence and uniqueness theorems for solutions (4.1) with and without jump components, are given in Gihman and Skorohod (1972). Although only scalar processes are considered in this chapter and the next chapter, the above vector formulation will be useful in Chapter 6 (Stochastic Partial Differential Equations).

4.3. The unrestricted membrane potential.

Before addressing the phenomenon of threshold crossings and neuronal firing, we determine some properties of the potential in the *absence* of a threshold. This is not without application for the following reason. When the drug tetrodotoxin (TTX) is applied to nerve cells, the sodium channels that are opened during an action potential are blocked. A neuron cannot then emit an action potential. However, the potassium channels are not blocked, and neither are the channels involved in synaptic transmission. Thus with random input to a cell treated with TTX, the voltage will be unrestricted except (locally) by reversal potentials for synaptic action, which do not enter the present model.

Before proceeding, we generalize the above description of Stein's model to allow for an arbitrary distribution of postsynaptic potential amplitudes. Then the stochastic differential equation for $V(t)$ is

$$(4.5) \qquad dV(t) = -V(t)\,dt + \int uN(dt, du),$$

where $\Lambda(.)$ is as yet unspecified. However, we assume the jump amplitude has a density so that we may write

$$\Lambda(du) = \phi(u)\,du.$$

Notice that time is now being measured in units of the time constant.

4.3.1. Moments. Let $V(0) = x$ almost surely. The mean and variance of $V(t)$ in this linear model can be found easily from the Green's function representation

$$V(t) = xe^{-t} + \int_0^t e^{-(t-t')} \int uN(dt', du).$$

This immediately yields

$$E(V(t)) = xe^{-t} + \mu_1(1 - e^{-t}),$$

where

$$\mu_1 = \int u\phi(u)\, du.$$

The variance is

$$\text{Var}(V(t)) = e^{-2t} \int_0^t e^{2t'} \int u^2 \phi(u)\, du\, dt'$$

$$= \frac{\mu_2}{2}(1 - e^{-2t}),$$

where

$$\mu_2 = \int u^2 \phi(u)\, du.$$

4.3.2. Transition density. For the stochastic equation (4.5) we obtain from (4.4) the following forward equation satisfied by the transition density:

$$\frac{\partial p}{\partial t} = \frac{\partial}{\partial y}(yp) - \lambda p + \int p(y - u, t \mid x, s)\phi(u)\, du.$$

This equation was considered prior to neural modeling in electric circuit theory by Keilson and Mermin (1959). The Fourier transform of p, which is the characteristic function of $V(t)$, is (dropping reference to x, s)

$$\bar{p}(k, t) = \int e^{iky} p(y, t)\, dy.$$

This is found to satisfy the first-order equation

$$\frac{\partial \bar{p}}{\partial t} = -k \frac{\partial \bar{p}}{\partial k} + \bar{p}(\bar{\phi} - \lambda),$$

where $\bar{\phi}$ is the Fourier transform of ϕ. Hence, it is found that

$$\bar{p}(k, t) = \exp\left\{\int_{ke^{-t}}^{k} \left(\frac{\bar{\phi}(k') - \lambda}{k'}\right) dk'\right\} \bar{p}_0(ke^{-t}),$$

where $\bar{p}_0(k)$ is the Fourier transform of the initial distribution. The transform of the steady-state distribution is thus

$$\bar{p}(k, \infty) = \exp\left\{\int_0^{k} \left(\frac{\bar{\phi}(k') - \lambda}{k'}\right) dk'\right\}.$$

In the case originally considered by Stein, we may take

$$\phi(u) = \lambda_E \delta(u - a_E) + \lambda_I \delta(u + a_I),$$

so we have $\lambda = \lambda_E + \lambda_I$, and the mean and variance of $V(t)$ are

$$(\lambda_E a_E - \lambda_I a_I)(1 - e^{-t}) \quad \text{and} \quad \tfrac{1}{2}(\lambda_E a_E^2 + \lambda_I a_I^2)(1 - e^{-2t}),$$

respectively.

A reasonable choice is

$$\phi(u) = c\,|u|\,e^{-a|u|},$$

in which case

$$\tilde{\phi}(k) = \frac{2c(a^2 - k^2)}{(a^2 + k^2)^2}$$

and the integrals in the expression for $\bar{p}$ can be evaluated in closed form.

4.4. First exit times and neuronal firing.

In many neuronal models the time taken to elicit an action potential is given by the time of first exit of a random process from a set in the phase space. Let A be such a set, and let $\{\mathbf{X}(t)\}$ be a temporally homogeneous process. Then the random variable of interest is

$$T_A(\mathbf{x}) = \inf\{t : \mathbf{X}(t) \notin A \mid \mathbf{X}(0) = \mathbf{x} \in A\},$$

which is the *first exit time* from A. If the indicated set is null, then we set $T_A(\mathbf{x}) = \infty$. When $\mathbf{X}(t)$ is a temporally homogeneous Markov process defined by a stochastic differential equation

$$(4.6) \quad d\mathbf{X}(t) = \boldsymbol{\alpha}(\mathbf{X}(t))\,dt + \boldsymbol{\beta}(\mathbf{X}(t))\,d\mathbf{W}(t) + \int \boldsymbol{\gamma}(\mathbf{X}(t), \mathbf{u})N(dt, d\mathbf{u}),$$

and with infinitesimal generator $\mathscr{A}$, we have the following result.

THEOREM 4.1. *Let $\mathbf{X}(t)$ be a Markov process satisfying (4.6), and assume that existence and uniqueness conditions are fulfilled. Then the distribution function*

$$F_A(\mathbf{x}, t) = \Pr\{T_A(\mathbf{x}) \le t\}$$

satisfies

$$(4.7) \quad \frac{\partial F_A}{\partial t} = (\mathscr{A} F_A(\cdot, t))(\mathbf{x}), \quad \mathbf{x} \in A, \quad t > 0,$$

with initial condition

$$F_A(\mathbf{x}, 0) = \begin{cases} 0, & \mathbf{x} \in A \\ 1, & \mathbf{x} \notin A, \end{cases}$$

and with boundary condition

$$F_A(\mathbf{x}, t) = 1, \quad \mathbf{x} \notin A, \quad t \ge 0.$$

A similar equation is satisfied by the density $f_A(\mathbf{x}, t)$ of $T_A(\mathbf{x})$ if this exists.

This result was proved for scalar processes in Tuckwell (1976a), and the proof given there extends easily to the vector case. In addition, we have the following integrodifferential equations for the moments of the first exit time.

COROLLARY 4.2. *If the moments*

$$\mu_n(\mathbf{x}) = E[T_A^n(\mathbf{x})], \quad n = 1, 2, \cdots,$$

exist, they satisfy the recursive system of equations

$$(4.8) \quad (\mathscr{A}\mu_n)(\mathbf{x}) = -n\mu_{n-1}(\mathbf{x}), \quad \mathbf{x} \in A$$

with boundary conditions

$$\mu_n(\mathbf{x}) = 0, \quad \mathbf{x} \notin A.$$

The quantity $\mu_0(\mathbf{x})$ is the probability that $\mathbf{X}(t)$ exits from A in a finite time and satisfies the equation

(4.9) $$(\mathcal{A}\mu_0)(\mathbf{x}) = 0, \quad \mathbf{x} \in A$$

with

$$\mu_0(\mathbf{x}) = 1, \quad \mathbf{x} \notin A.$$

In connection with μ_0 we have the following useful lemma, due to Gihman and Skorohod (1972).

LEMMA 4.3. *If there exists a bounded function $g(\mathbf{x})$ satisfying*

(4.10) $$(\mathcal{A}g)(\mathbf{x}) \leq -1, \quad \mathbf{x} \in A,$$

then $\mu_1(\mathbf{x}) < \infty$ and $\Pr\{T_A(\mathbf{x}) < \infty\} = 1$.

Equations (4.8) are a generalization of some earlier results of Darling and Siegert (1957) for scalar diffusion processes.

4.4.1. Results for Stein's model. Applying Lemma 4.3 to the scalar process satisfying (4.6) with $\alpha(x) = -x$, $\beta = 0$, $\gamma(x, u) = u$, $\phi(u) = \lambda_E \delta(u - a_E) + \lambda_I \delta(u + a_I)$, $A = (-\infty, \theta)$, we can deduce that the Stein model neuron fires in a finite time with probability one and with finite mean interspike interval. This follows because the solution of (4.9) is $\mu_0(x) = 1$, and this satisfies (4.10).

The mean first passage time to θ for an initial value x satisfies, from (4.8),

$$-x\frac{d\mu_1(x)}{dx} + \lambda_E \mu_1(x + a_E) + \lambda_I \mu_1(x - a_I) - (\lambda_E + \lambda_I)\mu_1(x) = -1, \quad x < \theta$$

with boundary condition $\mu_1(x) = 0$ for $x \geq \theta$. This differential-difference equation is difficult to solve, but some progress has been made (Cope and Tuckwell (1979)). When there is excitation so that $\lambda_I = 0$, analytical expressions for μ_1 can be obtained for small θ. The solutions are of interest in themselves.

4.4.2. Excitation only. For simplicity, let us set $\lambda_E = a_E = 1$ and $\theta = 2$ so that we need to solve, replacing μ_1 by f,

(4.11) $$-xf'(x) + f(x+1) - f(x) = -1.$$

We distinguish the following two cases for the interval I on which this equation holds.

Case (i): $I = (\epsilon, 2)$, $0 < \epsilon < 1$. Using the boundary condition at $x \geq 2$ and letting the solution on $[1, 2)$ be $f_1(x)$, we obtain, upon integrating the resulting differential equation,

$$f_1(x) = 1 + c_1/x.$$

Letting the solution on $(\epsilon, 1)$ be f_2, we have (using f_1 on this interval)

$$f_2(x) = 2 + [c_1 \ln(1+x) + c_2]/x.$$

Using the boundary condition $f_2(\epsilon) = 0$ and continuity on $(\epsilon, 2)$, we obtain the following values for the constants:

$$c_1 = \frac{1 - 2\epsilon}{1 + \ln[(1 + \epsilon)/2]},$$

$$c_2 = \frac{2\epsilon(\ln 2 - 1) - \ln(1 + \epsilon)}{1 + \ln[(1 + \epsilon)/2]}.$$

Case (ii): $I = [0, 2)$. The point $x = 0$ is special since if $V(0) = 0$, we wait on average for a time interval of length 1 ($\equiv 1/\lambda_E$) for the first jump. Consistent with the differential-difference equation, this gives

$$f(0) = 1 + f(1).$$

Solving as in Case (i) on $(\epsilon, 2)$, and letting $\epsilon \to 0$ together with the requirement of boundedness, we get $c_2 = 0$ and $c_1 = 1/(1 - \ln 2)$. For the initially resting neuron ($x = 0$), this yields

$$E[T] = 2 + \frac{1}{1 - \ln 2}.$$

Some observations of interest include the following: (a) if there were no decay $E[T] = 2$. The extra term $(1 - \ln 2)^{-1}$ reflects the contributions from the decay between EPSPs; (b) if the input was deterministic with the same value as the mean for the Poisson case so that $dV/dt = -V + 1$, the neuron would never fire because as $t \to \infty$, $V(t) \to 1$; (c) as $\epsilon \to 0$, the solution of (4.11) approaches that on $[0, 2)$. There is a boundary layer effect near $x = 0$.

Several exact results for the mean firing time of a Stein model neuron with excitation have been obtained. From Tuckwell (1975), for example, we have that when $\theta = 2$, $\lambda_E = n$ a positive integer,

$$E[T] = \frac{1}{n} \left[2 + \frac{1}{1 - n\left\{\ln 2 + \sum_{k=1}^{n-1} C_{nk}(1 - 2^{-k})\right\}} \right],$$

where

$$C_{nk} = \frac{(-)^k}{k} \binom{n-1}{k},$$

and when $\theta = 2$, $\lambda_E = 1/n$,

$$E[T] = n \left[2 + \frac{1}{1 - \dfrac{1}{n!} \sum_{k=0}^{\infty} \dfrac{(-)^k \Gamma\left(k + \dfrac{1}{n}\right)}{\left(k + \dfrac{1}{n}\right) k!}} \right],$$

which reduces when $n = 2$ to

$$E[T] = 2\left[2 + \frac{1}{1 - \ln(1 + \sqrt{2})} \right].$$

Further results were presented in Tuckwell (1976b). Using parameter values appropriate for dorsal spinocerebellar tract (DSCT) cells, a family of output frequency versus input frequency curves were obtained for various θ. Such curves were approximately linear over wide ranges. DSCT cells are also discussed in §4.5.

In addition to coinciding with a model for integrated shot noise (cf. Keilson and Mermin (1959)), Stein's model is the same as a model for the content of a dam, in which case the first passage time corresponds to the time to overflow. Results obtained by Yeo (1974) in this latter context using a different method agreed with those of Tuckwell (1976b), with which a comparison is possible. Yeo (1974), Losev (1975), and Tsurui and Osaki (1976) obtained explicit formulas for the moments of the first passage time to a constant threshold when the jump amplitude has an exponential distribution. The mean is given by

$$E[T] = \frac{1}{\lambda} + \Gamma(\lambda) \sum_{n=1}^{\infty} \frac{\tilde{\theta}^n}{n\Gamma(\lambda+n)},$$

where time is in time constants and $\tilde{\theta}$ is the ratio of threshold to the mean jump size.

It is natural to ask whether Stein's model, with its neglect of many anatomical and physiological details, can predict neuronal firing patterns. In one experiment, cat spinal motoneurons were monosynaptically excited according to a Poisson process (Redman, Lampard, and Annal (1968)). The EPSP amplitudes in this experiment were large at about one-third threshold. However, Stein's model could not predict the variation in mean output frequency as mean input frequency increased (Tuckwell (1976c)). It is not known which of many possible inhibiting influences might also be operating, but Tuckwell (1978a) concluded that the most likely one was recurrent inhibition via Renshaw cells. Partial confirmation of Stein's model has come from the fact that parameters estimated for real neurons (such as time constants, ratio of threshold to EPSP amplitude, and input frequency) have been plausible (Losev, Shik, and Yagodnitsyn (1975); Tuckwell and Richter (1978)). In the latter reference, a method was given for numerically solving, in conjunction with (4.11), the equations

$$-xg'(x) + g(x+1) - g(x) = -2f(x), \qquad -xh'(x) + h(x+1) - h(x) = -3g(x),$$

where $g(x)$ and $h(x)$ are the second and third moments of the first exit time. Exact results were also obtained for the variance for small θ and hence for the dependence of coefficient of variation of the interspike interval on mean interval. This revealed interesting structure not apparent in the computer simulation results of Stein (1965).

4.4.3. Excitation and inhibition. When there is inhibition present as well as excitation, the differential-difference equations (such as the one given earlier for $\mu_1(x)$) are extremely difficult to solve. An approximate method of solution was first given when the inhibition was weak relative to the excitation

(Tuckwell (1975)). In Cope and Tuckwell (1979) a method of solution was given that gave accurate results where a comparison was possible with results known exactly (excitation only) or from computer simulations. An asymptotic formula was obtained for values of x much less than θ; the solution was continued to values of x near θ and the boundary condition employed to find unknown constants.

4.4.4. Time-varying thresholds. As pointed out above, there is a refractory period immediately following a spike which can be understood after considering the solitary wave trajectories of solutions of the Hodgkin–Huxley equations. To incorporate this refractoriness into one-dimensional models, the threshold for firing may be elevated just after an action potential. In an absolute refractory period $\theta = \infty$, after which θ is usually taken to be a monotonically decreasing function of time. A method of solving first exit time problems for a class of moving barriers was developed by Tuckwell (1981). Let $\{X(t)\}$ be a random process satisfying the scalar version of the stochastic differential equation (4.6) with $X(0) = x$; and let $Y(t)$, $t \geq 0$, taken to be a deterministic function of time, satisfy a differential equation

$$\frac{dY}{dt} = \phi(Y),$$

with $Y(0) = y$. To be specific, assume that $x < y$ and that one required the time $T(x, y)$ of first passage of $X(t)$ to be barrier $Y(t)$. The idea is to consider the vector $(X(t), Y(t))$ as a partially degenerate vector-valued Markov process so that $T(x, y)$ can be interpreted as the first exit time of the process $\mathbf{X}(t) = (X(t), Y(t))$ from the region $y > x$. The infinitesimal generator $\mathscr{A}_{XY}$ of $\mathbf{X}(t)$ is given by

(4.12)
$$(\mathscr{A}_{XY} f)(x, y) = \alpha(x) \frac{\partial f}{\partial x} + \frac{\beta^2(x)}{2} \frac{\partial^2 f}{\partial x^2}$$
$$+ \int [f(x + \gamma(x, u), y) - f(x, y)] \Lambda(du) + \phi(y) \frac{\partial f}{\partial y}.$$

The following is actually a restatement of Corollary 4.2.

COROLLARY 4.4. *If the nth moments $\mu_n(x, y)$ of $T(x, y)$ exist, then they satisfy*

(4.13) $$(\mathscr{A}_{XY} \mu_n)(x, y) = -n\mu_{n-1}(x, y), \qquad x < y$$

with boundary conditions

$$\mu_n(x, y) = 0, \qquad x \geq y.$$

Similar equations apply to the distribution function and, if it exists, the density of $T(x, y)$ (Tuckwell and Wan (1984)). An application of this result is discussed in the next chapter. To conclude this section, we mention the reviews of Blake and Lindsey (1973) and of Abrahams (1985) on first passage time problems.

4.5. Extensions and generalizations.

One easily introduced modification to the stochastic equation (4.5) is to add a *steady current* to get

(4.14) $$dV(t) = (c - V(t))\,dt + \int uN(dt, du).$$

If $c > \theta$, the cell will fire in the absence of any random input at regular intervals of length

(4.15) $$t = \ln\left(\frac{c - V_0}{c - \theta}\right),$$

where the initial potential is $V_0 < \theta$. The equation for the mean first passage time to θ when the Poisson input is present becomes

$$(c - x)\frac{df}{dx} + f(x + 1) - f(x) = -1,$$

where it is assumed that $\lambda_E = a_E = 1$ and there is no inhibition. This equation may be solved as in the case $c = 0$.

When $c > \theta$, there is a background firing at a rate given by (4.15) plus noisy input. This was investigated as a model for DSCT cells by Walloe, Jansen, and Nygaard (1969). It could explain some, but not all, the features of the discharges of the neurons studied and was useful for other cells (Enger, Jansen, and Walloe (1969)). Note that the constant c in (4.14) may be negative, which leads to a decreased firing rate (cf. Tuckwell (1975)).

4.5.1. Inclusion of reversal potentials. When a synapse is activated, conductance increases occur for certain ion species. The changes in conductance for the various species (usually limited to Cl^-, K^+, Na^+) depend on the nature of the synapse and, more specifically, on the nature of the receptors. Increases in sodium conductance, for example, lead to excitatory effects.

The local current flow I_i at a synapse for ion species i is assumed to be

$$I_i = g_i(V_i - V),$$

where V_i is the corresponding Nernst potential, g_i is the conductance, and V the membrane potential. Since synaptic activation may lead to conductance changes for more than one ion species, a net *reversal potential* (i.e., a potential at which the current is zero and changes sign as V passes through it) is usually apparent. Such excitatory and inhibitory reversal potentials were found experimentally (see, for example, Eccles (1964)) and are usually denoted V_E, V_I, respectively. An expression for synaptic reversal potentials in terms of Nernst potentials is sometimes possible (Tuckwell (1985)).

Inclusion of the reversal potentials makes the nerve cell response to synaptic input depend on the state of the cell when the input occurs. With Poisson input activating short-lasting conductance changes, the following stochastic differential equation was employed by Tuckwell (1979):

(4.16) $\quad dV(t) = -\alpha V(t)\,dt + [V_E - V(t)]a_E\,dN_E(t) + [V_I - V(t)]a_I\,dN_I(t),$

where a_E, a_I are nonnegative constants, V_E, V_I are presumed to be constant, and usually we may expect $V_I \leq 0 < \theta < V_E$; α, N_E, N_I are as previously defined.

We may put (4.16) in the form of the general scalar stochastic differential equation (i.e., the scalar form of (4.6) by choosing $\alpha(X) = -\alpha X$, $\beta(X) = 0$, and

$$\gamma(X, u) = \begin{cases} (V_E - X)a_E, & u > 0, \\ (V_I - X)a_I, & u < 0, \end{cases}$$

with $\Lambda((-\infty, 0)) = \lambda_I$ and $\Lambda((0, \infty)) = \lambda_E$. Application of (4.2) gives the infinitesimal generator with action

(4.17) $\quad (\mathcal{A}f)(x) = -\alpha x \dfrac{df}{dx} + \lambda_E f(x + (V_E - x)a_E) + \lambda_I f(x + (V_i - x)a_I)$
$\qquad\qquad - (\lambda_E + \lambda_I)f(x).$

The forward Kolmogorov equation is obtained by utilizing

$$\gamma^*(z, u) = \begin{cases} \dfrac{a_E}{1 - a_E}(V_E - z), & u > 0, \\ \dfrac{a_I}{1 - a_I}(V_I - z), & u < 0, \end{cases}$$

which gives, from (4.4),

$$\dfrac{\partial p}{\partial t} = \alpha \dfrac{\partial}{\partial y}(yp)$$
$$+ \dfrac{\lambda_E}{1 - a_E} p\left(y - \dfrac{a_E(V_E - y)}{1 - a_E}, t\,|\,x, s\right)$$
$$+ \dfrac{\lambda_I}{1 - a_I} p\left(y - \dfrac{a_I(V_I - y)}{1 - a_I}, t\,|\,x, s\right) - (\lambda_E + \lambda_I)p(y, t\,|\,x, s).$$

The moments of the potential may be found from this equation or from the stochastic equation (4.16). For example, the mean depolarization in the absence of any threshold is given by

$$E(V(t)\,|\,V(0) = x) = \dfrac{k_2}{k_1} + \left(x - \dfrac{k_2}{k_1}\right)e^{-k_1 t},$$

where $k_1 = 1 + a_E\lambda_E + a_I\lambda_I$, $k_2 = a_E\lambda_E V_E + a_I\lambda_I V_I$. (See also Smith and Smith (1984).) Note that $V(t)$ is confined to the interval (V_I, V_E) if x is in this interval.

An exact result was obtained for the mean firing time with excitation only when the excitatory reversal potential is included (Tuckwell (1979)). This quantity satisfies $(\mathscr{A}f)(x) = -1$, where $\mathscr{A}$ is given by (4.17). In contrast to the case where reversal potentials are ignored, the intervals on which this differential-difference equation is solved by the method of steps are not all the same length. When we choose $\theta = a_E V_E(2 - a_E)$, there are two subintervals from 0 to θ; the mean firing time, ignoring refractoriness, is found to be

$$E[T] = 2 + \frac{(1 - a_E)}{[1 - a_E - \ln(2 - a_E)]}$$

where α has been set at unity and $\lambda_E = 1$. A comparison with results for Stein's model and computer simulations that are related to experimental results for cells in the cat cerebral cortex (Burns and Webb (1976)) can be found in Tuckwell (1978b), (1979). Subsequent analysis was performed by Wilbur and Rinzel (1982), (1983).

CHAPTER 5

One-Dimensional Diffusion Processes

The functional differential equations that arise from Stein's model are, as we saw in the previous chapter, difficult or cumbersome to solve. When the discontinuous Markov processes are replaced by approximating diffusions, the corresponding equations are differential equations for which there is a large body of literature in regard to both analytical and numerical methods of solution. In this chapter we introduce the well-known Wiener and Ornstein–Uhlenbeck processes as representatives of the state of a neuron. We also consider the less studied diffusion which arises when reversal potentials are employed, and we briefly consider the relevant weak convergence results which have only recently been obtained.

5.1. Diffusion processes.

A scalar diffusion process $\{X(t)\}$ has a stochastic differential equation of the form

$$dX(t) = \alpha(X(t), t)\, dt + \beta(X(t), t)\, dW(t),$$

where $\{W(t)\}$ is a standard Wiener process and α and β are real-valued functions of their arguments.

Conditions on α and β which guarantee that $\{X(t)\}$ is a diffusion process are given in Gihman and Skorohod (1972). As was pointed out in Chapter 4, in the definition of the solution

$$X(t) = X(s) + \int_s^t \alpha(X(\tau), \tau)\, d\tau + \int_s^t \beta(X(\tau), \tau)\, dW(\tau), \qquad t \geq s,$$

the stochastic integral is assumed to be defined in Ito's manner. The infinitesimal generator of the process in the *temporally homogeneous case* is then defined for suitable functions $f(x)$ by

(5.1) $$(\mathcal{A}f)(x) = \alpha(x)\frac{df}{dx} + \frac{1}{2}\beta^2(x)\frac{d^2f}{dx^2}.$$

Abbreviating $p(y, t\,|\,x, 0)$ to $p(y, t\,|\,x)$, we find that this transition density satisfies the *backward Kolmogorov equation*

$$\frac{\partial p}{\partial t} = \alpha(x)\frac{\partial p}{\partial x} + \frac{1}{2}\beta^2(x)\frac{\partial^2 p}{\partial x^2}$$

and the *forward Kolmogorov equation*

$$\frac{\partial p}{\partial t} = -\frac{\partial}{\partial y}(\alpha(y)p) + \frac{1}{2}\frac{\partial^2}{\partial y^2}(\beta^2(y)p).$$

It can be seen from the form of (5.1) that the equations (4.7) and (4.8) for the distribution function and moments of a first exit time become linear second-order differential equations.

5.2. Wiener process with drift.

The diffusion approximation to the randomized random walk is the Wiener process with drift. Gerstein and Mandelbrot (1964) turned to this approximation and compared the predicted firing time distributions with some experimental ones.

The process under consideration has the simple stochastic differential equation

$$dX(t) = \mu\, dt + \sigma\, dW(t),$$

where μ is a constant, called the drift, and σ is also a constant. (Of course, μ and σ could both depend on time, but we assume otherwise.) In order to see what μ and σ might signify, we generalize the randomized walk model slightly by putting

$$V(t) = a_E N_E(t) - a_I N_I(t).$$

Then, in the usual or standard diffusion approximation in which the approximating process and the original process have the same *infinitesimal mean*,

$$(5.2) \qquad \alpha(x, t) = \lim_{\Delta t \to 0} \frac{1}{\Delta t} E[X(t + \Delta t) - X(t) \mid X(t) = x],$$

and *infinitesimal variance*,

$$(5.3) \qquad \beta^2(x, t) = \lim_{\Delta t \to 0} \frac{1}{\Delta t} \text{Var}[X(t + \Delta t) - X(t) \mid X(t) = x],$$

we see that we require

$$(5.4) \qquad \mu = a_E \lambda_E - a_I \lambda_I$$

and

$$(5.5) \qquad \sigma^2 = a_E^2 \lambda_E + a_I^2 \lambda_I.$$

5.2.1. Transition density and first passage times.
It is immediate that the transition probability density is

$$p(y, t \mid x) = \frac{1}{\sqrt{2\pi\sigma^2 t}} \exp\left\{-\frac{(y - x - \mu t)^2}{2\sigma^2 t}\right\}, \quad x \in R, \quad t > 0,$$

because $X(t)$ is Gaussian with mean $x + \mu t$ and variance $\sigma^2 t$.

Assuming a constant threshold $\theta > x$, the density of the first passage time,

$$T_\theta(x) = \inf\{t : X(t) = \theta \mid X(0) = x\},$$

can be found in closed form. Although the method of images can be employed to this end in the case $\mu = 0$, a renewal equation approach works for all cases (summarized in Tuckwell (1988a)). The latter yields the following expression for the Laplace transform of the density $f_\theta(t; x)$ of $T_\theta(x)$:

$$f_{\theta, L}(s; x) = \exp\left\{\frac{\theta - x}{\sigma^2}\left(\mu - \sqrt{\mu^2 + 2\sigma^2 s}\right)\right\}.$$

The value of this at $s = 0$ enables us to find

$$\Pr\{T_\theta(x) < \infty\} = \begin{cases} 1, & \mu \geq 0, \\ \exp\left\{\frac{-2|\mu|(\theta - x)}{\sigma^2}\right\}, & \mu < 0. \end{cases}$$

That is, a neural firing is certain in a finite time if and only if $\mu \geq 0$. The inversion of $f_{\theta, L}$ gives the well-known expression

$$f_\theta(t; x) = \frac{\theta - x}{\sqrt{2\pi\sigma^2 t^3}} \exp\left\{-\frac{(\theta - x - \mu t)^2}{2\sigma^2 t}\right\}, \quad t > 0.$$

Of interest are the facts that the result for $\mu = 0$ was found by Bachelier in 1900 in connection with stock market fluctuations and that the general result was obtained by Schrödinger in 1915 in connection with physical Brownian motion. The result has come to be known as the *inverse Gaussian distribution*.

In terms of the original physiological parameters we have, when $x = 0$ and $\mu \geq 0$,

$$E(T_\theta) = \frac{\theta - x}{\mu} = \frac{\theta}{a_E\lambda_E - a_I\lambda_I}, \quad \text{Var}(T_\theta) = \frac{(\theta - x)\sigma^2}{\mu^3} = \frac{\theta(a_E^2\lambda_E + a_I^2\lambda_I)}{(a_E\lambda_E - a_I\lambda_I)^3},$$

these moments being infinite when $\mu = 0$. The *coefficient of variation* of the interspike interval (its mean divided by its standard deviation) is often employed to quantify the regularity of a spike train. In the present instance this is

$$CV = \frac{1}{\sigma}\sqrt{\frac{\mu}{\theta}},$$

where μ, σ are given by (5.4) and (5.5). Thus, for fixed values of μ and σ, the train of spikes is more regular the higher the threshold.

Despite the successful fitting of ISI histograms for the inverse Gaussian distribution (e.g., by Gerstein and Mandelbrot (1964)), the Wiener process with drift cannot be regarded as a valid model for nerve cell potential. The reason for this is that, as we have seen, there is exponential decay to resting value between inputs, and this exponential decay has a severe effect on the ISI. The results for the Wiener process with drift are of interest because they are the only ones available in closed form.

5.3. The Ornstein–Uhlenbeck Process (OUP).

This celebrated random process may be introduced in a manner similar to the Wiener process. Beginning with Stein's discontinuous model described in §4.1, we find the infinitesimal mean and variance using (5.2) and (5.3) to be

$$\alpha(x) = -\alpha x + a_E \lambda_E - a_I \lambda_I$$

and

$$\beta^2(x) = a_E^2 \lambda_E + a_I^2 \lambda_I.$$

This suggests a diffusion approximation whose stochastic differential equation is

(5.6) $$dX(t) = (-\alpha X(t) + \mu)\, dt + \sigma\, dW(t),$$

where μ and σ are as defined above. The corresponding forward Kolmogorov equation is

$$\frac{\partial p}{\partial t} = \frac{\partial}{\partial y}[(\alpha y - \mu)p] + \frac{\sigma^2}{2}\frac{\partial^2 p}{\partial y^2}.$$

This equation may be solved probabilistically as follows. Putting $Y(t) = X(t)e^{\alpha t}$ and applying Ito's formula gives

$$dY(t) = e^{\alpha t}[\mu\, dt + \sigma\, dW(t)].$$

Thus the mean and variance of $Y(t)$ may be easily found, and hence those of $X(t)$:

$$E[X(t)] = xe^{-\alpha t} + \frac{\mu}{\alpha}(1 - e^{-\alpha t}), \qquad \text{Var}[X(t)] = \frac{\sigma^2}{2\alpha}(1 - e^{-2\alpha t}).$$

Since $X(t)$ is a Gaussian random variable, we must have

(5.7) $$p(y, t \mid x) = \left(\frac{\alpha}{\pi\sigma^2(1 - \exp(-2\alpha t))}\right)^{1/2}$$
$$\times \exp\left\{\frac{-[y - xe^{-\alpha t} - \mu(1 - e^{-\alpha t})/\alpha]^2}{\sigma^2(1 - \exp(-2\alpha t))/\alpha}\right\},$$

where $-\infty < y < \infty$, as first given by Gluss (1967) in a neural context. Note that $\pm\infty$ are *natural boundaries* for the OUP, whereas all other points are regular (see, e.g., Kannan (1979) for Feller's boundary classification).

For the unrestricted OUP we may also obtain the covariance function. Defining this as $K(s, t) = \text{Cov}(X(s), X(t))$, $s \leq t$ it is readily seen that

$$K(s, t) = \frac{\sigma^2}{2\alpha}[e^{-\alpha(t-s)} - e^{-\alpha(t+s)}], \qquad s \leq t.$$

As $s, t \to \infty$, with $\tau = t - s$, this approaches

$$\tilde{K}(\tau) = \frac{\sigma^2}{2\alpha} e^{-\alpha\tau},$$

which means that for large times an OUP approaches a weakly stationary process. Since it is also a real Gaussian process, the limiting process is strongly stationary and we denote this by $\{\tilde{X}(t)\}$. Note that $\tilde{X}$ is the process that results for $t \geq 0$ when the initial distribution (i.e., of $X(0)$) is the stationary (Gaussian) distribution with mean μ/σ and variance $\sigma^2/2\alpha$. Before discussing first passage time theory for the OUP, we take a look at the ways in which neurophysiologists used it to account for ISI variability.

5.3.1. Applications.
Let us first point out that the OUP has arisen in many areas outside neurophysiology and that often the quantities sought have been similar (e.g., first passage times). It was introduced by Uhlenbeck and Ornstein (1930) to represent the speed of a particle undergoing (physical) Brownian motion. It appeared in astrophysics (Chandresekhar (1943)), electrical circuit theory (Stumpers (1950)), queueing theory (Iglehart (1965)), birth and death theory (McNeil and Schach (1973)), and population biology (Tuckwell (1974)).

Motoneuron firing patterns. Calvin and Stevens (1965), (1968) made a careful study of the randomness of the ISIs of cat spinal motoneurons using intracellular recording. The voltage fluctuations in the absence of stimulation were analyzed and found to have peak to peak amplitudes between 2 and 8 mV. Their amplitude distribution was found to be approximately Gaussian, and their autocovariance approximately an exponentially decreasing function of time difference with time constant of about 4 msec. It was inferred that the voltage fluctuations were due to randomly arriving postsynaptic potentials (EPSPs and IPSPs) and that they could be represented by a stationary OUP, $\tilde{X}(t)$, as defined above.

Observations on membrane potential trajectories during repetitive firing of these cells showed that the average terminal trend was approximately linear. The serial correlation between neighboring ISIs was found to be negligible so that a satisfactory stochastic model needed only to predict the ISI distribution (renewal process). For many, but not all cells, this distribution was well approximated by the first passage time of

$$V(t) = a + bt + \tilde{X}(t)$$

to a constant level θ, as was verified by computer simulation. It is possible that the stochastic differential equation for the depolarization

$$dV = \left(-\frac{V}{RC} + \frac{I}{C}\right) dt + \sigma\, dW$$

would be adequate here, where I is the injected current; R, C have their usual meanings; and W represents the random synaptic currents. Calvin and Stevens concluded that their model adequately predicted firing patterns for a class of neurons and that for them the *only* source of randomness in ISI was random PSPs.

Repetitive firing in superior olivary complex neurons. Geisler and Goldberg (1966) investigated the adequacy of a model similar in principle to Hagiwara's (1954). They wanted to predict the stochastic discharge patterns of neurons in the cat superior olivary complex in response to long-lasting acoustic stimuli; thus they set

$$V(t) = a + \tilde{X}(t),$$

where a is constant and $\{\tilde{X}(t)\}$ is a stationary OUP. The threshold function was

$$\theta(t) = \begin{cases} \infty, & 0 \le t \le t_R, \\ b + \dfrac{e^{-k(t-t_R)}}{1 - e^{-k(t-t_R)}}, & t > t_R, \end{cases}$$

where $t_R \cong 0.7$ msec is the length of the absolutely refractory period, and b and k are constants. After a spike the threshold was reset and the value of $\tilde{X}(t)$ chosen arbitrarily, presumably according to the stationary density (see below for further discussion). The stochastic behavior of many superior olive neurons was successfully duplicated by simulations of the above model. An extended model, which included afterhyperpolarization, was able to account for negative serial correlation found in some cells.

On the use of stationary processes. The above model of Calvin and Stevens can be presented as follows. $V(t)$ is the sum of a stationary OUP $\tilde{X}(t)$ and a time-varying deterministic function $a + bt$. An action potential occurs when $V(t)$ reaches the constant θ. Thus a neuron firing occurs when $\tilde{X}(t)$ hits the linearly decreasing barrier $\theta - (a + bt)$. The question arises as to what happens just subsequent to a threshold crossing. We may assume that in this model, as in Hagiwara's (see Chapter 3), the *threshold* is reset, but not the random component of the voltage; if the latter occurs, then the process will no longer be stationary.

We thus see that the point models with threshold-crossing phenomena separate into two classes: the type used by Gerstein and Mandelbrot and by Stein, in which the voltage and threshold are reset after a spike; and the kind employed by Hagiwara, by Calvin and Stevens, and by Geisler and Goldberg, in which a random voltage component is not reset but the threshold is (notwithstanding the fact that all of the second group were considering a stimulus plus noise). However, the difference in ISIs predicted by the two model classes often will be small, and they will be significant only if the neuron under consideration is firing rapidly. The reason for this is that the time constant of approach to stationarity is the membrane time constant, which, for spinal motoneurons, for example, is about 4–5 msec. If the initial voltage is chosen from the stationary density, however, there is a nonzero chance that the voltage will be above threshold at the beginning of the ISI. This is clearly undesirable.

In connection with stationary processes, there are some results that may be useful in the present context (see, e.g., Cramer and Leadbetter (1967)). Perhaps we need reminding that white-noise-driven systems are only an idealization so that even though the following does not apply to the OUP it may apply to certain physiological processes.

Let $\{X(t)\}$ be a real-valued, Gaussian stationary process with mean μ, covariance function $r(\tau)$, $\tau \geq 0$, and spectral function $F(.)$:

$$r(\tau) = \int_0^\infty \cos(\lambda \tau)\, dF(\lambda).$$

Let the $2k$th moment of F, if it exists, be

$$\lambda_{2k} = \int_0^\infty \lambda^{2k}\, dF(\lambda), \quad k = 0, 1, \cdots,$$

and suppose further that X has continuous sample paths with probability one. Then the expected number of upcrossings of level θ per unit time is

$$E(N_\theta) = \frac{1}{2\pi} \left(\frac{\lambda_2}{\lambda_0}\right)^{1/2} \exp\{-(\theta - \mu)^2 / 2\lambda_0\}.$$

This predicts a dramatic drop in firing frequency as $\theta - \mu$ increases.

Finally, we point out a possible relation between the models that employ a stationary random process for the voltage and those that assume the voltage starts at $x < \theta$, θ constant, leading to an expected ISI $E(T_\theta | x)$. If the voltage does start above θ, we put the firing time at t_R and start the process again. Then

$$E(T_\theta) = t_R \int_\theta^\infty p_{\text{st.}}(x)\, dx + \int_{-\infty}^\theta (t_R + E(T_\theta | x)) p_{\text{st.}}(x)\, dx$$

$$= t_R + \int_{-\infty}^\theta E(t_\theta | x) p_{\text{st.}}(x)\, dx,$$

where $p_{\text{st.}}$ is the stationary density.

5.3.2. First passage times for the OUP. The earliest results on first passage times for the OUP were those of Chandresekhar (1943), Wang and Uhlenbeck (1945), and Darling and Siegert (1953). There has been little progress with analytical methods of solution since then, which explains why researchers have often resorted to simulations. Gluss (1967) considered the unrestricted OUP, Johannesma (1968) seems to have first looked at first passage time theory in a neural setting, and Roy and Smith (1969) first obtained an expression for the mean firing time. The following first exit time theory is a standard application of the theory outlined in §4.4.

With time in units of the membrane time constant, if $\{X(t), t \geq 0\}$ is an OUP with $X(0) = x$, then the density $f_{ab}(x, t)$ of the first exit time from (a, b) satisfies

$$\frac{\partial f_{ab}}{\partial t} = (\mu - x)\frac{\partial f_{ab}}{\partial x} + \frac{\sigma^2}{2}\frac{\partial^2 f_{ab}}{\partial x^2}, \quad t > 0, \quad x \in (a, b),$$

with boundary conditions

$$f_{ab}(a, t) = f_{ab}(b, t) = \delta(t), \quad t \geq 0,$$
$$f(x, 0) = 0, \quad a < x < b.$$

The best attack has proven to be by Laplace transforms. Put

$$f_{ab,L}(x, s) = \int_0^\infty e^{-st} f_{ab}(x, t)\, dt,$$

to get

$$\frac{\sigma^2}{2} f''_{ab,L} + (\mu - x) f'_{ab,L} - s f_{ab,L} = 0, \quad x \in (a, b),$$

with boundary conditions

$$f_{ab,L}(a, s) = f_{ab,L}(b, s) = 1.$$

Of direct interest is the first passage time to level $\theta > 0$ for an initial value $x < \theta$. The corresponding density $f_\theta(x, t)$ is obtained from

$$f_\theta(x, t) = \lim_{a \to \infty} f_{-a, \theta}(x, t), \quad a > 0.$$

Moments. From §4.4 we find that the moments $\mu_n(x)$ of the first exit time from $(-a, \theta)$ satisfy

(5.8) $$\frac{\sigma^2}{2}\mu''_n + (\mu - x)\mu'_n = -n\mu_{n-1}, \quad x \in (-a, \theta).$$

When $n = 0$, we get the probability of escape from $(-a, \theta)$. Solving the equation for μ_0 with boundary conditions

$$\mu_0(-a) = 0, \quad \mu_0(\theta) = 1,$$

gives

$$\mu_0 = \frac{\int_{-a}^x \exp\{2(\frac{y^2}{2} - \mu y)/\sigma^2\}\, dy}{\int_{-a}^\theta \exp\{2(\frac{y^2}{2} - \mu y)/\sigma^2\}\, dy}.$$

As $a \to \infty$, $\mu_0(x) \to 1$. Hence, denoting the firing time by T_θ,

$$\Pr\{T_\theta < \infty\} = 1,$$

for all μ and all σ.

Roy and Smith obtained the following expression for the Laplace transform of the first passage time density to level θ:

$$f_{\theta,L}(x, s) = \frac{\Psi\left(\frac{s}{2}, \frac{1}{2}; \left(\frac{\mu-x}{\sigma}\right)^2\right)}{\Psi\left(\frac{s}{2}, \frac{1}{2}; \left(\frac{\mu-\theta}{\sigma}\right)^2\right)},$$

where $\Psi(.,.;.)$ is a confluent hypergeometric function of the second kind (see, e.g., Abramowitz and Stegun (1965)). (See also Capocelli and Ricciardi (1971).)

The mean firing time, assuming an initially resting cell, is $\mu_1(0)$, obtained from the solution of (5.8) with boundary conditions $\mu_1(-a) = \mu_1(\theta) = 0$ and by letting $a \to \infty$. The following result was also obtained by Roy and Smith:

(5.9)
$$E(T_\theta) = \sum_{k=0}^{\infty} \frac{2^k}{(2k+1)!! (k+1)} (Y^{2k+2} - Z^{2k+2})$$
$$+ 2\sqrt{\pi}\left[Z\Phi\left(\frac{1}{2}, \frac{3}{2}; Z^2\right) - Y\Phi\left(\frac{1}{2}, \frac{3}{2}; Y^2\right)\right],$$

where

$$Y = (\mu - \theta)/\sigma, \qquad Z = \mu/\sigma,$$

and $\Phi(.,.;.)$ is the confluent hypergeometric function of the first kind. For computational purposes, (5.9) is efficient for moderate firing rates; otherwise, numerical integration of the differential equation may be a better method, or evaluation of solutions by quadratures (see, e.g., Thomas (1975)).

Higher-order moments of T_θ have also been found. The *second moment* was found in series form by Ricciardi and Sacerdote (1979). Expressions from which the *third moment* could also be calculated were given by Sato (1978).

Given that there are several parameters, the presentation of computed results for the OUP is cumbersome. First passage time densities and its first two moments were *tabulated* in some cases by Keilson and Ross (1975). Analytical approximate results are thus useful. Using integral representations for the mean first passage time, Thomas (1975) developed accurate *approximations* involving Dawson's integral.

By scaling, the first passage time problem for the OUP can be reduced to that of $X(t)$, satisfying

$$dX(t) = (\eta - X(t))\,dt + \epsilon\,dW(t),$$

to unity, that is, put $\eta = \mu/\theta$ and $\epsilon = \sigma/\theta$. When $\epsilon^2 \ll 1$, perturbation methods were used by Wan and Tuckwell (1982) to obtain approximate solutions of the differential equations satisfied by the moments of time the T of first reaching threshold. Three cases were distinguished, depending on the relative magnitudes of the mean steady-state value of X (namely, η) and the threshold (1).

Case (a). This case involves a mean steady-state response greater than threshold with

$$\epsilon \ll \eta - 1.$$

Then asymptotically as $\epsilon \to 0$, we have

$$E(T) \sim \ln\left(\frac{\eta}{\eta-1}\right) - \frac{\epsilon^2}{4}\left[\frac{1}{(\eta-1)^2} - \frac{1}{\eta^2}\right],$$

$$\text{Var}(T) \sim \frac{\epsilon^2}{2}\left[\frac{1}{(\eta-1)^2} - \frac{1}{\eta^2}\right].$$

Case (b). This case involves a mean steady-state response less than threshold with

$$\epsilon \ll 1 - \eta.$$

Then

$$E(T) \sim \frac{\epsilon\sqrt{\pi}}{1-\eta} e^{(1-\eta)^2/\epsilon^2},$$

$$\text{Var}(T) \sim \frac{\epsilon^2 \pi}{(1-\eta)^2} e^{2(1-\eta)^2/\epsilon^2}.$$

Case (c). This case involves a mean steady-state response near threshold with

$$|1 - \eta| = O(\epsilon).$$

The expressions obtained in this case can be found in Wan and Tuckwell (1982). One conclusion in that work was that the effect of noise is to decrease the expected firing time, even if the noise has zero mean. This was not obvious from the above formula of Roy and Smith, and it explains some experimental findings on crayfish stretch receptors (Buno, Fuentes, and Segundo (1978)). The coefficient of variation of T was investigated in great detail in Wan and Tuckwell (1982) with a view to ascertaining the following: (i) when the output train of spikes is more regular than the input train; and (ii) the significance of a coefficient of variation greater than unity (cf. Tuckwell (1979)).

Time-dependent thresholds. In a special case, the density of the first passage time of an OUP may be found in closed form. If $X(t)$ satisfies (5.6) and we put (Doob (1942))

$$\tau = e^{2\alpha t}, \qquad U(\tau) = \left(X(t) - \frac{\mu}{\alpha}\right)e^{\alpha t},$$

then $U(\tau) = \sigma W(\tau)/\sqrt{2\alpha}$, where W is a standard Wiener process. Now $X(t)$ hits the barrier $\theta(t) = \zeta + \xi e^{-\alpha t}$, where ζ and ξ are constants, when $U(\tau)$ hits the barrier $v(\tau) = \xi + (\zeta - (\mu/\alpha))\sqrt{\tau}$. Thus results may be obtained for the OUP in the case of a constant barrier from those for a Wiener process with square root boundary. Furthermore, in the special case

$$\zeta = \frac{\mu}{\tau},$$

the first passage time problem for the OUP is reduced to that of a (driftless) Wiener process to a constant barrier. The latter problem has an exact solution, as given in §5.2, and this may be suitably transformed to give a special case solution for the OUP (Sugiyama, Moore, and Perkel (1970)). These authors, incidentally, also presented a firing time density calculated numerically by the Crank–Nicolson method.

Mean firing intervals of a model neuron with threshold functions satisfying first-order differential equations may be found numerically by applying Corollary 4.4. For example, if the barrier $Y(t)$ satisfies

$$\frac{dY}{dt} = -k_1 Y + k_2, \quad t > 0, \quad Y(0) = y,$$

then the mean time $F(x, y)$ at which $X(t)$ satisfying (5.6), with initial value x, hits $Y(t)$, can be found by solving

$$(\mu - \alpha x)\frac{\partial F}{\partial x} + (k_2 - k_1 y)\frac{\partial F}{\partial y} + \frac{\sigma^2}{2}\frac{\partial^2 F}{\partial x^2} = -1, \quad x < y$$

with $F(y, y) = 0$. Some numerically obtained solutions were given for the exponentially decaying threshold case in Tuckwell and Wan (1984).

Comparison with the discontinuous model. The OUP obtained with μ and σ given by (5.4) and (5.5) has the same infinitesimal mean and variance as the original discontinuous process it is being used to approximate. In fact, both processes have the same mean, variance, and covariance function. It is of interest, therefore, to compare the firing times predicted by the two models. Using results from Cope and Tuckwell (1979), it was possible to do this via the mean firing time because the results for the OUP were readily obtainable from Roy and Smith's formula (5.9). The values of a_E and a_I were held at unity, and the relationship between the predictions of the two models obtained as functions of λ_E for various values of λ_I and θ. The limiting values for both models as $\lambda_E \to \infty$ were known, as was the relative error in using the diffusion approximation in the limit $\lambda_E \to 0$. From these various pieces of information, the general scheme relating the two sets of results was deduced without excessive computation. A full account is given in Tuckwell and Cope (1980).

5.4. Inclusion of reversal potentials.

A major deficiency of the above diffusion models is that there is nothing to stop the membrane potential from making excursions to extremely hyperpolarized states, which would not arise in the course of synaptic inhibition. This is because, as we have seen in §4.5, reversal potentials limit such excursions, although of course a cell may be hyperpolarized by injected current beyond V_I. Gerstein and Mandelbrot (1964) tried to overcome this difficulty with a reflecting barrier at a hyperpolarized level.

A diffusion process based on the discontinuous process introduced by

Tuckwell (1979) may be obtained by employing the first two infinitesimal moments of the discontinuous process as those of the diffusion. This gives the following stochastic differential equation for subthreshold potentials:

$$(5.10) \quad \begin{aligned} dX = &[I - X + a_E \lambda_E (V_E - X) + a_I \lambda_I (V_I - X)] \, dt \\ &+ [a_E^2 \lambda_E (V_E - X)^2 + a_I^2 \lambda_I (V_I - X)^2]^{1/2} \, dW, \end{aligned}$$

where we have also introduced an injected current of scaled magnitude I.

The use of this diffusion approximation introduces a complication in that with $I = 0$, the original discontinuous process, in the absence of a threshold for firing, was confined to the interval (V_I, V_E). Clearly, $\{X(t)\}$ defined by (5.10) is not so confined. This may be ignored without much error but can also be remedied by the placement of reflecting barriers at the reversal potentials. When there is excitation only, zero is accessible to the diffusion approximation but not to the original discontinuous process if $X(0) \in (0, V_E)$.

The diffusion (5.10) was investigated by Hanson and Tuckwell (1983). The mean, variance, and stationary distribution for (5.10) were found, in the absence of threshold for excitation only, both on $[0, V_E)$ and on $(-\infty, V_E)$, depending on the boundary conditions. The same quantities were found for excitation and inhibition, both on (V_I, V_E) and on $(-\infty, \infty)$. Mean firing times were calculated from the differential equation $(\mathcal{A}\mu_1)(x) = 1$, where $\mathcal{A}$ is the infinitesimal generator of X defined in (5.10). These calculations, which are quite involved, agreed fairly closely with previous results obtained by simulation with the discontinuous model. (The reader may consult Hanson and Tuckwell (1983) for further details.)

5.5. Weak convergence results.

We expect diffusion approximations to the discontinuous processes described in Chapters 3 and 4 to perform best, in some sense, when the original processes have small, frequent jumps. Justification of the diffusion approximations, therefore, was originally sought by showing that the infinitesimal moments of all orders (and, hence, the Kolmogorov equations) of a sequence of discontinuous processes became, in the limits of vanishing postsynaptic amplitudes and infinite input frequencies, the corresponding quantities for diffusion processes (Roy and Smith (1969), Capocelli and Ricciardi (1971), Tuckwell and Cope (1980)). Such a justification by itself, however, did not yield satisfactory information about the relation between the processes in the sequence and the limiting process. More precise statements can be made in terms of the *weak convergence* of a sequence of processes to a limiting process. Such an approach has been useful in other areas, such as genetics (see, for example, Ethier and Kurtz (1986, Chap. 10)) and has recently been pursued in the present context. A clear account of weak convergence is given in Billingsley (1968). Here we give a brief summary of some of the basic ideas and the results that have been obtained for neural models.

The discontinuous processes employed above to represent the subthreshold membrane potentials of neurons have sample paths in the space $D[0, \infty)$

(henceforth abbreviated "D"), the space of real-valued functions on $[0, \infty)$ which at each point are right-continuous and have left-hand limits. A certain metric, the Skorohod metric, makes D a complete, separable metric space.

Let $(\Omega, \mathscr{F}, P)$ be a suitable probability space. A map $X: \Omega \to D$ defines a random function whose realizations (corresponding to points ω in Ω) constitute sample paths of a (possibly) discontinuous process. The induced probability measure P_X, defined on Borel sets of D by

$$P_X(\Gamma) = P\{X(\omega) \in \Gamma\}, \qquad \Gamma \in \mathscr{B}(D),$$

is called the distribution of X. Naturally, $P_X(D) = 1$.

At fixed $t \in [0, \infty)$, we obtain a random variable (R-valued) which may be denoted by $X_t(\omega)$ or just X_t—but we use the notation $X(t)$. The collection $\{X(t), t \geq 0\}$ is an R-valued random process for which results are obtained by working with the D-valued random variable X.

A sequence of probability measures $\{P_n\}$ on $\mathscr{B}(D)$ is said to converge weakly to P if $P_n(A) \to P(A)$ for each $A \in \mathscr{B}(D)$ such that $P(\partial A) = 0$. This is often stated as the equivalent condition that $\int f\, dP_n \to \int f\, dP$ for all bounded continuous real functions f on D. If the $\{P_n\}$ and P are the distributions of $\{X_n\}$ and X, we say the sequence $\{X_n\}$ converges in distribution to X. The notations for these statements are $P_n \xrightarrow{W} P$ and $X_n \xrightarrow{\mathscr{D}} X$, respectively.

One method of establishing weak convergence involves tightness. A probability measure P on $(D, \mathscr{B}(D))$ is said to be tight if for each $\epsilon > 0$ there exists a compact set $K \in \mathscr{B}(D)$ such that $P(K^c) < \epsilon$, where K^c is the complement of K. A sequence $\{P_n\}$ is called *tight* if P_n is tight for each n. It can be shown that under the Skorohod metric, a sequence $\{P_n\}$ converges weakly to P if, essentially,

(a) $\{P_n\}$ is tight; and

(b) The finite-dimensional distributions of $\{P_n\}$ converge to those of P. (See Billingsley (1968, Thm. 15.1), for complete details.)

In the neuronal models, we are concerned with the convergence of the discontinuous processes to limiting diffusions. Technique for establishing such weak convergence results include convergence of generators (Kurtz (1981)) and convergence of sequences of semimartingales (Rebolledo (1979), Liptser and Shiryayev (1984)). These may be applied to obtain sequences of randomized random walks converging weakly to a Wiener process (Lánský, (1984)), sequences of discontinuous processes with exponential decay converging weakly to an Ornstein–Uhlenbeck process (Kallianpur (1983), Lánský (1984)), and convergence of discontinuous processes with exponential decay and reversal potentials converging weakly to a diffusion process of the type defined in (5.10) (Lánský and Lánská (1987), Kallianpur and Wolpert (1987)). It was also shown that the first passage times to a constant threshold for sequences of Stein-type models converged in distribution to that of an Ornstein–Uhlenbeck process (Kallianpur (1983), Lánský (1984)). It has not yet been indicated, however, how these results can be applied to determine approximately the firing time distributions of a (model) neuron which receives excitation and inhibition of given amplitudes and given frequencies.

CHAPTER 6

Stochastic Partial Differential Equations

The neuronal models examined in Chapters 3–5 are often called *point models* to indicate that the neurons they represent are regarded as lumped circuits. In §3.2 we saw that one method of arriving at such models was to average over space in a model which incorporated the spatial extent of a cell. There is no doubt that such an approach leads to a great mathematical simplification, for we then deal only with one-dimensional random processes. When these are Markov, the theory is worked out except for numerical solutions of the many associated differential equations and other equations. Of course, the assumptions that lead to Markovity (white noises, Poisson processes) are only approximately valid even in the best of circumstances in a controlled laboratory experiment.

Such one-dimensional models are expected to yield only gross details of neuronal input-output relations. The application to motoneuron firing with injected current and synaptic noise was successful (Calvin and Stevens (1968)), but application to motoneuron firing with Poisson stimulation (Redman, Lampard, and Annal (1968)) indicated that there were complicating factors.

Deterministic neural modeling had taken into account the spatial extent of a nerve fiber by using cable theory in one space dimension. This was extended, as we saw in Chapter 1, to the treatment of dendritic trees by the use of various mapping procedures. The advantages of such models, which include variation in space of *neuronal dynamical variables,* are, apart from the obviously greater physical realism, the following:

(i) The *location of a given input,* excitatory or inhibitory, with given temporal characteristics, can be taken into account. Since impulsive conductance changes remote from the soma give rise to smoothly rising postsynaptic potentials (PSPs) at the soma in the cable model, we automatically take care of the problem of PSP shapes (cf. White and Ellias (1979)).

(ii) *Different locations* for multiple inputs can be taken into account and their interactions and joint effects studied.

(iii) Various *temporal sequences* of inputs at various locations can be incorporated.

We remark that (ii) and (iii) are expected to be important in the study of receptive fields, direction-sensitive cells, etc. (see, for example, Koch, Poggio, and Torre (1983), for a deterministic study). In an early study, Fernald (1971) introduced Poisson noise in an analogue circuit model in order to predict poststimulus time histograms of cat cochlear nucleus neurons.

Our immediate program is as follows. We will first formulate a general neuronal model of cable-type employing a lumped soma boundary condition at the origin to represent a neuron with dendritic trees emanating from a common soma. The cases of linear stochastic partial differential equations (SPDEs) involving white noise inputs will be considered, including results for special cases obtained thus far, such as white noise at a point or over a patch and two-parameter white noise. We will then turn to nonlinear SPDEs such as those of Hodgkin and Huxley with noise, examine previous modeling, and introduce a perturbative approach which involves Green's function matrices. In addition, we will consider the space-clamped versions which lead to, for example, multidimensional diffusion processes. We will see that the transition from point models to those that include spatial extent involves a very large increase in mathematical complexity and, when the analysis is performed rigorously, a concomitant increase in mathematical sophistication.

6.1. General cable models for neurons with synaptic and injected input currents.

We will begin by considering a single nerve cylinder of length L, possibly, if certain geometrical constraints are satisfied, representing a dendritic tree. Rall's lumped R–C circuit representation is used for the soma at $x = 0$. We allow for the fact that there may be several different kinds of synapse at the same distance from the soma and that various synapses when activated involve conductance increases for possibly several ion species. Then the potential (depolarization) along the cylinder must approximately satisfy the following PDE (Tuckwell (1985, eq. (1.10)), with t and x dimensionless,

$$(6.1) \quad V_t = V_{xx} - V + \sum_{i=1}^{n} g_i(x, t)(V_i - V) + r_m I_A(x, t), \quad 0 < x < L,$$

where the subscripts t and x denote partial differentiation with respect to these variables, r_m is the membrane resistance of unit length times unit length, and $I_A(x, t)$ is the applied current density through nonsynaptic sources. The summation is over different ionic species, and these would usually include chloride, potassium, and sodium with perhaps calcium and other cations. The quantity g_i is the ratio of the *increase* in conductance per unit length for ionic species i to the total resting membrane conductance per unit length. This makes $V = 0$ an equilibrium state when there are no synapses activated and when there is no applied current. The Nernst potentials V_i (relative to resting level) are assumed to be constant, but under certain conditions, such as extremely rapid firing, will fluctuate in accordance with departures of the ionic concentrations from their usual values. This complication will be ignored.

The boundary condition on (6.1) at $x = L$ depends on the experimental conditions. Usually, if this point represents a neuron terminal, then a zero longitudinal current condition will be appropriate:

(6.2) $$V_x(L, t) = 0.$$

A more complicated boundary condition is needed at $x = 0$ if a lumped soma is assumed there. With a slight extension of the usual form of this condition, we obtain

(6.3) $$\frac{R_s I_s(t)}{\tau} + \sum_{i=1}^{n} g_{s,i}(t)(V_i - V(0, t)) = V(0, t) + \sigma V_t(0, t) - \gamma V_x(0, t),$$

where R_s is the resting soma resistance, τ is the cylinder membrane time constant, $I_s(t)$ is current injected at the soma, $g_{s,i}(t)$ is the ratio of the conductance increase for ionic species i to the resting soma conductance, γ is the ratio of the resting soma resistance to the resistance of a characteristic length of the adjacent nerve cylinder, and σ is the ratio of the soma time constant to that of the cylinder.

Any of the currents I_s and I_A and any of the conductance terms g_i and $g_{s,i}$, $i = 1, \cdots, n$, may be random. If $I_s(t)$ is a one-parameter white noise, then we are dealing with an experiment in which white noise current is injected into the soma. If the cell is receiving random synaptic input, then the quantities $g_i(x, t)$ and $g_{s,i}(t)$ are random processes. A random I_A may represent nonsynaptic channel noise. In all these cases, $\{V(x, t), 0 \leq x \leq L, t \geq 0\}$ is a two-parameter random process or *random field* and satisfies a stochastic partial differential equation.

In some circumstances it is possible to describe the input currents due to synaptic activation in terms of reversal potentials. Using an obvious notation, (6.1) then becomes

(6.4) $$V_t = V_{xx} - V + \sum_{i=1}^{m} g_{E,i}(x, t)(V_{E,i} - V) + \sum_{j=1}^{n} g_{I,j}(x, t)(V_{I,j} - V) + r_m I_A,$$

where we have distinguished excitatory and inhibitory contributions, $V_{E,i}$ and $V_{I,i}$ being synaptic reversal potentials, and allowed for several kinds of excitatory and inhibitory endings on the same neuron. Similarly, the quantities $g_{s,i}$ and V_i in (6.3) may be replaced by ones specific for synaptic rather than ionic properties.

6.1.1. A neuron with arbitrary dendritic geometry. Equations (6.1) or (6.3) apply to a single cylinder of uniform radius. Let us consider a neuron with a soma, axon, and several dendritic trees. The latter are composed of several *dendritic segments*, which for simplicity we assume are of uniform radius, although including a varying radius is not difficult. We may distinguish the following four kinds of dendritic segment:

  (i) One end adjoins the soma, the other end is a dendritic terminal;
  (ii) One end adjoins the soma, the other end is connected to other dendritic

segments;

(iii) Both ends are connected to other dendritic segments;

(iv) One end is connected to other dendritic segments; the other end is a dendritic terminal.

The boundary conditions for each kind will be different. For any ending joining the soma, an equation like (6.3), or its modified version involving synaptic quantities, will apply, whereas at dendritic terminals, we may use (6.2) or its generalization to allow for leakage,

$$c_1 V(L, t) + c_2 V_x(L, t) = 0,$$

where c_1 and c_2 are constants. The only other type of condition needed is that where a dendritic segment of length L', say, for which we will use the subscript 1 and assume a parental role, adjoins daughter cylinders $2, \cdots, M$. At such junctions we impose the constraints of continuity of electrical, potential

$$V_1(L', t) = V_2(0, t) = \cdots = V_M(0, t),$$

and conservation of axial current,

$$\frac{V_{1,x}(L'^-, t)}{\bar{r}_1} = \sum_{k=2}^{M} \frac{V_{k,x}(0^+, t)}{\bar{r}_k},$$

where $\bar{r}_k$ is the internal resistance of a characteristic length of segment k. Thus we obtain a possibly large number of coupled SPDEs whose analysis is not tractable. Hence, we have the utility of approximations or mapping theorems which greatly reduce the number of equations. Such simplifications will be made in the next few sections.

6.2. A Rall model neuron with random input current.

In the Rall model neuron, devised to approximate the neurophysiology of the cat spinal motoneuron, the equivalent cylinder is employed to represent the dendrites and a lumped R–C circuit represents the soma. The quantity σ in (6.3) is usually set to unity; rather than employ Nernst potentials as reversal potentials, synaptic inputs are represented by applied currents, as in the terms I_s or I_A in (6.3). There are many forms of input that we could take, but we will choose just one here that corresponds to an actual neurophysiological experiment and discuss the others elsewhere.

The experiment to which we refer is the injection of white noise current into a neuron's soma (e.g., Moore and Christensen (1985)). We employ the following equations:

(6.5a) $\qquad V(x, 0) = 0,$

(6.5b) $\qquad V(0, t) + V_t(0, t) - \gamma V_x(0, t) = R_S\left(\alpha + \beta \dfrac{dW}{dt}\right),$

(6.5c) $\qquad V_t = V_{xx} - V, \quad 0 < x < L,$

(6.5d) $\qquad V_x(L, t) = 0.$

The meaning of this system is given in terms of an integral representation of its solution by means of the Green's function G^*, which has been given for this problem in Bluman and Tuckwell (1987). Thus

$$V(x, t) = \alpha \int_0^t G^*(x, t - u) \, du + \beta \int_0^t G^*(x, t - u) \, dW(u),$$

where the integral with respect to W is a Wiener or Ito integral. Since G^* has an eigenfunction expansion

$$G^*(x, t) = \sum_{n=0}^{\infty} \bar{A}_n(x) e^{-\mu_n t},$$

we see that at fixed x, $V(x, t)$ is the *sum of an infinite number of Ornstein–Uhlenbeck processes*

$$V(x, t) = \sum_{n=0}^{\infty} V_n(x; t),$$

where the V_n's satisfy the ordinary stochastic differential equations

$$dV_n = (\alpha \bar{A}_n(x) - \mu_n V_n) \, dt + \beta \bar{A}_n(x) \, dW.$$

The following results for the moments readily follow. For the mean,

(6.6)
$$E(V(x, t)) = \alpha \int_0^t G^*(x, t - u) \, du$$

$$= E(V(x, \infty)) - \alpha \bar{r}_i \sum_{n=0}^{\infty} \frac{A_n(x) e^{-(\lambda_n^2 + 1)t}}{\lambda_n^2 + 1},$$

where the steady-state mean is

(6.7)
$$E(V(x, \infty)) = \frac{\alpha \bar{r}_i \cosh(L - x)}{(1/\gamma + \tanh L) \cosh L},$$

λ_n, $n = 0, 1, \cdots$ are the roots of

$$\gamma \tan(\lambda L) + \lambda = 0,$$

$$A_0(x) = \frac{\gamma}{1 + \gamma L},$$

$$A_n(x) = \frac{2\gamma \cos(\lambda_n (L - x))}{\cos(\lambda_n L)\{1 + \gamma L\} - \lambda_n L \gamma \sin(\lambda_n L)}, \quad n = 1, 2, \cdots,$$

and $\bar{r}_i$ is the internal cylinder resistance per unit characteristic length. The covariance of the voltage at two space-time points is

(6.8)
$$\operatorname{Cov}(V(x, s), V(y, t)) = \beta^2 \int_0^s G^*(x, s - u) G^*(y, t - u) \, du, \quad s \leq t,$$

$$= (\beta \bar{r}_i)^2 \sum_{m=0}^{\infty} \sum_{n=0}^{\infty} \frac{A_m(x) A_n(y)}{\lambda_m^2 + \lambda_n^2 + 2} [\exp\{(\lambda_n^2 + 1)(s - t)\}$$
$$- \exp\{-(\lambda_m^2 + 1)s - (\lambda_n^2 + 1)t\}].$$

Hence, the variance is given by

$$\text{Var}(V(x,t)) = (\beta\bar{r}_i)^2 \sum_{m=0}^{\infty} \sum_{n=0}^{\infty} \frac{A_m(x)A_n(x)}{\lambda_m^2 + \lambda_n^2 + 2}[1 - \exp\{-(\lambda_m^2 + \lambda_n^2 + 2)t\}],$$

which clearly approaches a constant as $t \to \infty$. Also, if we put $t - s = T$, then the covariance function becomes asymptotically, at large times,

(6.9) $$K(x, y; T) = (\beta\bar{r}_i)^2 \sum_{m=0}^{\infty} \sum_{n=0}^{\infty} \left(\frac{A_m(x)A_n(y)}{\lambda_m^2 + \lambda_n^2 + 2}\right)\exp\{-(\lambda_n^2 + 1)T\}.$$

From these facts and the fact that V is obtained by linear operations on a Gaussian process, we get the following theorem.

THEOREM 6.1. $\{V(x, t)\}$ at fixed x is a Gaussian process with mean given by (6.6) and covariance function given by (6.8) with $x = y$. As time becomes infinite, this process asymptotes to a strongly stationary Gaussian process with mean given by (6.7) and covariance function given by (6.9) with $x = y$.

6.3. Synaptic input and other boundary conditions.

Consider a nerve cylinder on $[a, b]$. In order to incorporate reversal potentials and randomly arriving synaptic conductance changes, we proceed as follows. To simplify the formalism, we assume there is only one kind of excitatory synapse and one kind of inhibitory synapse, with reversal potentials V_E, V_I, respectively. Let N_E and N_I be independent random counting measures on Borel sets of $[0, \infty) \times [0, \infty) \times [a, b]$. Then

$$N_E(A \times B \times C) = \int_A \int_B \int_C N(dt, du, dx)$$

is the number of excitatory synaptic conductance changes in the time subset A with amplitudes in $B \in \mathcal{B}((0, \infty))$ in the set $C \in \mathcal{B}([a, b])$ along the length of the cylinder. This is also true for the inhibitory inputs. The SPDE describing the evolution of the depolarization is then

(6.10)
$$\frac{\partial V}{\partial t} = \frac{\partial^2 V}{\partial x^2} - V + (V_E - V)\frac{\partial^2}{\partial x \, \partial t}\left[\int_0^t \int_0^\infty \int_0^x u' N_E(dt', du', dx')\right]$$
$$+ (V_I - V)\frac{\partial^2}{\partial x \, \partial t}\left[\int_0^t \int_0^\infty \int_0^x u' N_I(dt', du', dx')\right]$$

The solution of this equation can be obtained quite straightforwardly using the Green's function for the regular cable equation and the given boundary conditions (see, e.g., Tuckwell (1985), (1986a)). Synaptic input may also be applied at the soma in the previous boundary condition.

In some circumstances we may assume that

$$\tilde{N}_E(t, B \times C) = N_E((0, t] \times B \times C), \quad t \geq 0,$$

is a Poisson process with rate $\Lambda_E(t, B \times C)$, and similarly for $\tilde{N}_I(t, B \times C)$.

In the temporally homogeneous case,

(6.11) $$E[\tilde{N}_E(t, B \times C)] = t\tilde{\Lambda}_E(B \times C),$$

and similarly for $\tilde{N}_I$. We would then have the cable analogue of the one-dimensional discontinuous Markov model

$$dV = -V\,dt + (V - V_E)\int_{R^+} uN(dt, du) + (V - V_I)\int_{R^-} uN(dt, du).$$

In the case where an excitatory synaptic input occurs at x_E and an inhibitory one occurs at x_I, we have, in the temporally homogeneous Poisson case,

$$\tilde{\Lambda}_{E,I}(du, dx) = \delta(x - x_{E,I})\,dx\,\tilde{\Lambda}_{E,I}(du)$$

so that we have points stimulated by compound Poisson processes

$$\frac{\partial V}{\partial t} = \frac{\partial^2 V}{\partial x^2} - V + \delta(x - x_E)(V_E - V)\frac{d}{dt}\int_0^t\int_0^\infty u'\tilde{N}_E(dt', du')$$

$$+ \delta(x - x_I)(V_I - V)\frac{d}{dt}\int_0^t\int_0^\infty u'\tilde{N}_I(dt', du').$$

Specializing further, we put

$$\tilde{\Lambda}_{E,I}(du) = \lambda_{E,I}\delta(u - a_{E,I}), \qquad a_{E,I} > 0$$

so that we get simply

$$\frac{\partial V}{\partial t} = \frac{\partial^2 V}{\partial x^2} - V + a_E\delta(x - x_E)(V_E - V)\frac{d\hat{N}_E}{dt} + a_I\delta(x - x_I)(V_I - V)\frac{d\hat{N}_I}{dt},$$

where $\hat{N}_{E,I}$ are independent simple Poisson processes with intensities $\lambda_{E,I}$. If we ignore reversal potentials, we obtain the cable analogue of Stein's model:

$$\frac{\partial V}{\partial t} = \frac{\partial^2 V}{\partial x^2} - V + a_E\delta(x - x_E)\frac{d\hat{N}_E}{dt} - a_I\delta(x - x_I)\frac{d\hat{N}_I}{dt}.$$

When $x_E = x_I = x_0$, we may obtain a diffusion analogue of this:

(6.12) $$\frac{\partial V}{\partial t} = \frac{\partial^2 V}{\partial x^2} - V + \left(\alpha + \beta\frac{dW}{dt}\right)\delta(x - x_0),$$

which was considered in Wan and Tuckwell (1979) and with white noise applied over an extended space region in Tuckwell and Wan (1980).

Assuming boundary conditions have been chosen so that a Green's function can be found, the solution of (6.12) is defined as

$$V(x, t) = \int_a^b G(x, y; t)V(y, 0)\,dy$$

$$+ \int_a^b\int_0^t G(x, y; t - s)[\alpha\,ds + \beta\,dW(s)]\,\delta(y - x_0)\,dy,$$

for $a \leq x \leq b$ and $t \geq 0$. Assuming the cell is initially at rest, we may set $V(x, 0) = 0$, $x \in [a, b]$. Then the y integration gives

$$V(x, t) = \alpha \int_0^t G(x, x_0; t-s) \, ds + \beta \int_0^t G(x, x_0; t-s) \, dW(s).$$

This decomposes V into a deterministic part equal to its mean

$$\mu(x, t) = E(V(x, t)) = \alpha \int_0^t G(x, x_0; t-s) \, ds$$

and a random part involving a one-dimensional stochastic integral. The covariance for general linear boundary conditions is

$$\mathrm{Cov}(V(x, s), V(y, t)) = \beta^2 \int_0^{s \wedge t} G(x, x_0; s-y) G(y, x_0; t-u) \, du.$$

Theorem 6.1 also applies in this case with appropriate changes in the first- and second-order moments and the proviso that $x \neq x_0$.

6.3.1. Examples. Although neurons are finite, the assumption of an infinite spatial domain is the only way to get closed form expressions for several quantities of interest.

(i) *Infinite cable*: $x \in (-\infty, \infty)$. The Green's function is

$$G(x, y; t) = \frac{\exp\{-t - (x-y)^2/4t\}}{\sqrt{4\pi t}}, \qquad t > 0.$$

Defining

$$\mathrm{erfc}(z) = \frac{2}{\sqrt{\pi}} \int_z^\infty e^{-x^2 dx},$$

we obtain

$$E(V(x, t)) = \frac{\alpha}{2\sqrt{2}} \left(\exp\{-|x-x_0|\} \mathrm{erfc}\left\{ \frac{|x-x_0| - 2t}{2\sqrt{t}} \right\} \right.$$
$$\left. - \exp\{|x-x_0|\} \mathrm{erfc}\left\{ \frac{|x-x_0| + 2t}{2\sqrt{t}} \right\} \right).$$

As $t \to \infty$, this gives a steady-state mean voltage peaked at the point of application of the current:

$$\mu(x, \infty) = \frac{\alpha}{\sqrt{2}} e^{-|x-x_0|}, \qquad x \in R.$$

The expression for the variance is somewhat unwieldy except in the steady state

$$\mathrm{Var}(V(x, \infty)) = \frac{\beta^2}{2\pi} K_0(2|x-x_0|), \qquad x \neq x_0,$$

where $K_0(x)$ is a modified Bessel function (see, e.g., Abramowitz and Stegun

(1965)). Note that $K_0(x)$ is singular at $x = 0$. We thus have, at fixed $x \neq x_0$,

$$V(x, t) \stackrel{d}{\underset{t \to \infty}{\sim}} N\left(\frac{\alpha}{\sqrt{2}} e^{-|x-x_0|}, \beta\left(\frac{K_0(2|x - x_0|)}{2\pi}\right)^{1/2}\right).$$

(ii) *Finite cable*: $x \in [0, L]$. Using an eigenfunction expansion for the Green's function

$$G(x, y; t) = \sum_{n=0}^{\infty} e^{-\mu_n^2 t} \phi_n(x) \phi_n(y), \qquad t > 0,$$

we find, as in §6.2, that at fixed x the depolarization is an infinite sum of Ornstein–Uhlenbeck processes. With Neumann conditions, which are the simplest with any degree of realism for a physically intact nerve cell, we have

$$\mu_n^2 = 1 + \frac{n^2 \pi^2}{L^2}, \qquad n = 0, 1, 2, \cdots$$

and the eigenfunctions are

$$\phi_0(x) = \frac{1}{\sqrt{L}}, \qquad \phi_n(x) = \sqrt{\frac{2}{L}} \cos\left(\frac{n\pi x}{L}\right).$$

(Note that for CNS neurons, L lies between about 0.5 and 2.0 for an equivalent cylinder.)

The mean is then

$$E(V(x, t)) = \alpha \sum_{n=0}^{\infty} \frac{[1 - \exp\{-\mu_n^2 t\}]}{\mu_n^2} \phi_n(x) \phi_n(x_0)$$

with steady-state value

$$E(V(x, \infty)) = \begin{cases} \dfrac{\alpha \cosh(L - x_0)\cosh x}{\sinh L}, & 0 < x \leq x_0, \\ \dfrac{\alpha \cosh x_0 \cosh (L - x)}{\sinh L}, & x_0 < x < L. \end{cases}$$

The variance of the voltage is

$$\text{Var}(V(x, t)) = \beta^2 \sum_{m=0}^{\infty} \sum_{n=0}^{\infty} \frac{[1 - \exp\{-(\mu_m^2 + \mu_n^2)t\}]}{\mu_m^2 + \mu_n^2}$$
$$\times \phi_m(x_0) \phi_m(x) \phi_n(x_0) \phi_n(x),$$

which converges if $x \neq x_0$. For further results and graphs, see Wan and Tuckwell (1979) and Tuckwell and Wan (1980).

At this point we make reference to some related works. The use of Green's functions to obtain solutions for linear SPDEs was elaborated on in Wan (1972). Becus (1977), (1978) obtained properties of the solutions of the random heat equation, which is essentially the same as the cable equation. Convergence of solutions of SPDEs with Poisson forcing terms to those driven by white noise was considered by Kallianpur and Wolpert (1984). Parabolic

SPDEs considered as stochastic equations for processes with values in a Hilbert space were studied by Curtain and Falb (1971) and Curtain (1977). Ito (1984) obtained several results for infinite dimensional Ornstein–Uhlenbeck processes.

6.4. First passage times.

For neuronal models involving linear SPDEs such as those in the previous three sections, there are several possible (ad hoc) threshold conditions that could be employed. To determine the ISI distribution theoretically, in the first instance we consider the following four first passage times to a constant or varying θ. For all of these we assume an initially resting cell so that $V(x, 0) = 0$ for all $x \in [0, L]$ and $\theta > 0$:

$$T_1 = \inf\left\{t : \frac{1}{L}\int_0^L V(x, t)\, dx \geq \theta\right\},$$

$$T_2 = \inf\left\{t : \sup_{0 \leq x \leq L} V(x, t) \geq \theta\right\},$$

$$T_3 = \inf\left\{t : V(x_1, t) \geq \theta\right\}, \qquad 0 \leq x_1 \leq L,$$

$$T_4 = \inf\{t : V(x, t) \geq \theta \;\forall x \in [x_1, x_2] \subseteq [0, L]\}.$$

The use of T_1 means that a spike will be emitted when the average depolarization reaches a certain level. When the Neumann boundary conditions are employed, this reduces to a first passage time problem for a one-dimensional Markov process and shows again a relation between the previous point models and the spatial models. If T_2 is used to find ISI distributions, then it is assumed that an action potential can be instigated when the depolarization at any point on the neuron reaches θ. Often cells have a region where the density of sodium channels is relatively high so that the threshold is lowest there. Such a region is called a "trigger zone" and its consideration motivates T_3 and T_4. For T_3 the trigger zone is considered to be just a single space point, whereas for T_4 the zone extends over the interval $[x_1, x_2]$. More generally, we could allow for a varying (in space) threshold over the neuron and put

$$T_5 = \inf\left\{t : \sup_{0 \leq x \leq L} (V(x, t) - \theta(x)) \geq 0\right\},$$

where $\theta(x)$ is usually increasing on $[0, L]$. The fact that we have so many choices for an ISI random variable draws attention to the deficiency of the use of linear equations with imposed threshold conditions to replace nonlinear systems with natural threshold properties (see §6.6). In our first investigations of this problem we employed T_1 with $x_1 = 0$, since this is one of the simplest and most reasonable approximations in many circumstances. It remains to be seen how different the ISI distributions are for the various possible threshold conditions.

6.4.1. Approximate solution.
Consider the cable equation

$$(6.13) \qquad V_t = V_{xx} - V + \sum_{m=1}^{M} [f_m(x)\{g_m(t) + \dot{X}_m(t)\}]$$

on a finite interval such that the Green's function takes the form

$$G(x, y; t) = \sum_n \phi_n(x)\psi_n(y) e^{-\mu_n t}, \qquad t > 0,$$

where $\{\phi_n\}$ are the spatial eigenfunctions, $\{\psi_n\}$ are the corresponding coefficients, and $\{\mu_n\}$ are the corresponding eigenvalues. In (6.13) there are M sources of synaptic input which may or may not be overlapping in space. The term $f_m(x)$ is nonrandom and describes the spatial extent of an input—a typical example would be unity in some bounded subinterval of the x domain of V and zero otherwise. The term $g_m(t)$ represents the mean or deterministic part of the temporal component of the noisy input and $\dot{X}_m(t)$ represents the random part, usually taken as Gaussian white noise or Poisson white noise. Employing the Green's function in the case of initially resting cell gives the representation

$$(6.14) \qquad V(x, t) = \sum_n \phi_n(x) \sum_{m=1}^{M} a_{mn} V_{mn}(t), \qquad t > 0,$$

where

$$a_{mn} = \int \psi_n(y) f_m(y)\, dy,$$

and V_{mn} is a one-dimensional process with stochastic differential equation

$$dV_{mn}(t) = (-\mu_n V_{mn} + g_m(t))\, dt + dX_m(t).$$

If the X_m's are Wiener processes, then each V_{mn} is an Ornstein–Uhlenbeck process; if the X_m's are derived from Poisson processes, then each V_{mn} is the kind of process in Stein's model. The approximation consists of considering only contributions to the sum in (6.14) up to, say, $n = N$. This makes it possible, when the g_m's are constant, at least, to use the standard theory of first exit times for vector-valued Markov processes that we gave in §4.4. If we designate the approximation by $V_N(x, t)$, and assume N terms, then the approximate firing time is (using the third of the possible firing time variables with $x_1 = 0$),

$$T_{3,N} = \inf\{t : V_N(0, t) \geq \theta\}$$
$$= \inf\left\{t : \sum_{n=1}^{N} \phi_n(x) \sum_{m=1}^{M} a_{mn} V_{mn}(t) \geq \theta\right\}.$$

This is a first exit time for an MN-dimensional process from a region in R^{MN}. In addition to using the analytical methods for Markov processes, we may also use *simulation*, which may be a superior alternative considering its simplicity versus the extremely large-order differential equations with which we are confronted in the former approach.

6.4.2. Example. As an illustration, consider the point white-noise-driven cable model neuron as in (6.12). As noted earlier, $V(x, t)$ is the sum of an infinite number of OUPs. Making a two-term approximation, we find that the mean ISI can be found by solving

$$(\alpha\phi_0(x_0)\phi_0(x) - \mu_0^2 x_1)\frac{\partial M}{\partial x_1} + (\alpha\phi_1(x_0)\phi_1(x) - \mu_1^2 x_2)\frac{\partial M}{\partial x_2}$$
$$+ \frac{\beta^2}{2}(\phi_0^2(x_0)\phi_0^2(x)\frac{\partial^2 M}{\partial x_1^2} + 2\phi_0(x_0)\phi_0(x)\phi_1(x_0)\phi_1(x)\frac{\partial^2 M}{\partial x_1 \partial x_2}$$
$$+ \phi_1^2(x_0)\phi_1^2(x)\frac{\partial^2 M}{\partial x_2^2} = -1,$$

the domain being $x_1 + x_2 < \theta$ and the boundary condition being $M = 0$ on $x_1 + x_2 = \theta$. Here the ϕ's are as given in §6.3, example (ii). In practice, one solves this partial differential equation on a finite rectangle with reflecting boundaries remote from $x_1 + x_2 = \theta$ and lets the rectangle grow until the solution in the region of interest is insensitive to further changes. The mean firing time in this approximation is $M(0, 0)$, and accurate values were found by finite-difference methods of solutions. For further details and results, see Tuckwell, Wan, and Wong (1984). In this reference, methods of simulation are also discussed and firing time distributions compared for various numbers of terms in the approximating sum. An interesting dependence of the shape of the ISI density was found upon the location of the synaptic input—an effect that would have been impossible to glean from the one-dimensional models of earlier chapters.

6.5. Linear cable with two-parameter white noise.

We have seen that at a given location (synapse) the randomness of an input stream may sometimes be approximated by a Poisson process. Suppose that the input channels, which may be synapses or just ion channels, are of uniformly high density in space. Then, if small equal spatial subintervals have equal chances to be the site of an input event at any given time, we expect the number of input events to be Poisson-distributed in space as well as in time. This motivates the use of two-parameter Poisson processes as driving terms for a cable equation. For example, if we have two such processes, N_E and N_I, homogeneous in space and time and independent of each other, then

(6.14a) $$\frac{\partial V}{\partial t} = \frac{\partial^2 V}{\partial x^2} - V + a_E \frac{\partial^2 N_E(x, t)}{\partial x \, \partial t} - a_I \frac{\partial^2 N_I(x, t)}{\partial x \, \partial t},$$

where $a_E, a_I > 0$ are constants and the intensities λ_E and λ_I of N_E, N_I are defined through

$$E[N_E(x, t)] = \lambda_E t \, |x - a|, \qquad E[N_I(x, t)] = \lambda_I t \, |x - a|,$$

where $x \in [a, b]$ and $t \geq 0$. The solution of (6.14a) is given, as usual, in terms of a Green's function, but we will not dwell on this process. Rather, we turn to a diffusion approximation.

The diffusion approximation here involves a two-parameter Wiener process $\{W(x, t), x \in [a, b], t \geq 0\}$. This is a mean zero Gaussian process (or random field) with covariance function

$$E[W(x, s)W(y, t)] = \min(x - a, y - a) \min(s, t).$$

Two-parameter white noise, which we denote by $\partial^2 W/\partial x \, \partial t$ or $w(x, t)$, can be viewed, as in the one-parameter case, as the formal derivative of W and the relation between them expressed by

$$\int_a^x \int_0^t w(y, s) \, dy \, ds = W(x, t).$$

The covariance of the white noise may be formally written

$$E[w(x, s)w(y, t)] = \delta(s - t) \, \delta(x - y).$$

Furthermore, stochastic integrals with respect to $W(x, t)$ or the equivalent Gaussian measure have been defined and their properties investigated (Park (1970), Zimmerman (1972), Wong and Zakai (1974)).

We are led, therefore, to consider the SPDE

(6.15) $$\frac{\partial V}{\partial t} = \frac{\partial^2 V}{\partial x^2} - V + \alpha + \beta \frac{\partial^2 W}{\partial x \, \partial t}, \quad a < x < b, \quad t \geq 0,$$

where $\alpha = \alpha(x, t)$ and $\beta = \beta(x, t)$ are nonrandom. Assuming smoothness requirements are met and V is initially zero, we define the solution of (6.15), as usual, in terms of the integral representation,

$$V(x, t) = \int_a^b \int_0^t \alpha(y, s) G(x, y; t - s) \, ds \, dy + \int_a^b \int_0^y \beta(y, s) G(x, y; t - s) \, dW(y, s).$$

Properties of V have been found in the case α, β constant by Tuckwell and Walsh (1983), although a similar equation was given as an example by Marcus (1974). It has also occurred in physics (Parisi and Wu (1981)) in the context of stochastic quantization. We will briefly summarize a few results.

In general we have for the *mean* voltage

$$E(V(x, t)) = \alpha \int_a^b \int_0^t G(x, y; t - s) \, ds \, dy,$$

and for the covariance with $s \leq t$,

$$\text{Cov}[V(x, s), V(y, t)] = \beta^2 \int_a^b \int_0^t G(x, z; s - u) G(y, z; t - u) \, du \, dz$$

$$= \tfrac{1}{2}\beta^2 \int_{t-s}^{t+s} G(x, y; u) \, du.$$

For an infinite cylinder, simple expressions are obtained for the following moments:

$$E(V(x, t)) = \alpha(1 - e^{-t})$$

$$\text{Var}(V(x, t)) = \frac{\beta^2}{4}(1 - \text{erfc}(\sqrt{2t})),$$

and the *spectral density* of the limiting (infinite time), resulting weakly stationary process at a given space point

$$f(\omega) = \frac{\beta^2}{4\pi\sqrt{2}} \cdot \frac{1}{\omega} \cdot \frac{\sqrt{(1+\omega^2)^{1/2} - 1}}{(1+\omega^2)^{1/2}},$$

where ω is the frequency. Of considerable interest is the large frequency behavior

$$f(\omega) \underset{\omega \to \infty}{\sim} \frac{\beta^2 \sqrt{2}}{8\pi} \omega^{-3/2}.$$

On bounded intervals, the potential is again an infinite series of Ornstein–Uhlenbeck processes. To illustrate, consider the case of *Neumann conditions* for a cable on $[0, L]$. Using the eigenfunction expansion for $G(x, y; t)$ and defining the following:

$$\alpha_n = \alpha \int_0^L \phi_n(y)\, dy, \qquad W_n(t) = \int_0^L \int_0^t \phi_n(y)\, dW(y, s),$$

so that $\{W_n(t), t \geq 0\}$ is a standard Wiener process and W_n, W_m are independent when $m \neq n$, we obtain the representation

$$V(x, t) = \sum_{n=0}^{\infty} \phi_n(x) V_n(t),$$

where the V_n's are independent OUPs with stochastic differential equations

$$dV_n = (\alpha_n - \lambda_n V_n)\, dt + \beta\, dW_n,$$

with $V_n(0) = 0$. Similar expansions hold for other boundary conditions.

In the present case, $(V_x(0, t) = V_x(L, t) = 0)$, some of the results obtained are the mean and variance with their steady-state values

$$E(V(x, t)) = \alpha(1 - e^{-t}) \underset{t \to \infty}{\longrightarrow} \alpha,$$

$$\mathrm{Var}(V(x, t)) = \beta^2 \sum_n \frac{\phi_n^2(x)}{2\lambda_n} [1 - e^{-2\lambda_n t}] \underset{t \to \infty}{\longrightarrow} \frac{\beta^2 \cosh(L-x) \cosh X}{2 \sinh L}.$$

As $t \to \infty$ at fixed x, we obtain a stationary Gaussian process. The spectral density of this process has been given for each x in Tuckwell and Walsh (1983). At the end of the cable, for example,

$$f(\omega; x = 0 \text{ or } L) = \frac{\beta^2}{4\pi\omega(1+\omega^2)^{1/2}} \left[\frac{\gamma \sinh(2L\rho) + \rho \sinh(2L\gamma)}{\gamma \sinh^2(L\gamma) + \sin^2(L\rho)} \right]$$

$$\underset{\omega \to \infty}{\sim} \mathrm{const.}\, \omega^{-3/2},$$

where

$$\gamma = \left(\frac{\sqrt{1+\omega^2} + 1}{2} \right)^{1/2}, \qquad \rho = \left(\frac{\sqrt{1+\omega^2} - 1}{2} \right)^{1/2}.$$

In Tuckwell and Walsh (1983), there is also an investigation into first passage times, and Walsh (1981) contains many technical results.

6.6. Systems of nonlinear stochastic ordinary and partial differential equations.

In 1952, Hodgkin and Huxley (HH) devised, on the basis of empirical evidence, a system of four equations which enabled us to predict a wide variety of phenomena involving nerve membrane. These equations were nonlinear, and analysis on them did not commence until many years later. The task of using the HH or similar equations on a whole neuron surface was intractable, so the cable approximation, valid for not-too-large subthreshold stimuli, was employed and culminated in Rall's model, which is much used by physiologists (see, e.g., Jack and Redman (1971), Finkel and Redman (1983)).

When stochastic neural modeling began in earnest in the 1960s, Rall's model was just being established mainly for deterministic studies on motoneurons. Not surprisingly, the simplest models were considered first, namely, the one-dimensional Markov models discussed in Chapters 3, 4, and 5. Since 1979, considerable effort has been put into studying solutions of the linear SPDEs that result from stochastic cable theory. Here, however, there are still several unaddressed problems, and to make satisfactory progress with them will probably take many more years.

The final step, in the context of models of single neurons, is the study of stochastic versions of nonlinear systems of equations such as those of Hodgkin and Huxley. There were some earlier approximate and simulation studies. The two-dimensional (V, m) approximation to the HH model due to Fitzhugh (1961) was employed by Lecar and Nossal (1971) to predict results such as those of Pecher (1939) on the probabilities of an action potential. A system of two ordinary stochastic differential equations involving additive white noise was analyzed and an expression obtained for the probability of an action potential for a given step change in voltage. Agreement with this experiment was certainly strong. Clay (1976) addressed the same problem using a random version of the linearized ordinary Fitzhugh–Nagumo (Bonhoeffer-van der Pol) equations (see below). The effects of randomly opening and closing sodium and potassium channels on solutions of the HH equations in both space-clamped (see below) and nonspace-clamped case were studied by using simulation, by Skaugen (1978). These were found to have a smoothing effect on the input-output relation.

In this section we will focus on two approaches to analyzing the properties of nonlinear random equations such as the HH equations. In the first, we consider the associated ordinary differential equations (Tuckwell (1986b)), and in the second we use a perturbative approach (Tuckwell (1987a), (1988c)) to obtain a solution of the full HH system with additive noise.

There are three systems of equations often employed in the modeling of neuronal activity. These are the HH equations, the Fitzhugh–Nagumo (Nagumo, Arimoto, and Yoshizama (1962)) equations, and the Frankenhaeuser–

Huxley equations (Frankenhaeuser and Huxley (1964)), which were devised to model the electrophysiological behavior at frog nodes of Ranvier. These systems are of dimension four, two, and five, respectively. We will consider only the first and second of these systems here.

Stochastic Hodgkin–Huxley equations. The HH equations may be written

$$C_m \frac{\partial V}{\partial t} = \frac{a'}{2\rho_i} \frac{\partial^2 V}{\partial x^2} + \bar{g}_K n^4 (V_K - V) + \bar{g}_{Na} m^3 h (V_{Na} - V)$$

$$+ g_l(V_l - V) + I(x, t), \qquad a < x < b,$$

$$\frac{\partial n}{\partial t} = \alpha_n(V)(1 - n) - \beta_n(V) n,$$

$$\frac{\partial m}{\partial t} = \alpha_m(V)(1 - m) - \beta_m(V) m,$$

$$\frac{\partial h}{\partial t} = \alpha_h(V)(1 - h) - \beta_h(V) h,$$

where C_m = membrane capacitance per unit area; a' = nerve fiber radius; ρ_i is the intracellular resistivity; $\bar{g}_K$ = constant maximal available potassium conductance per unit area; V_K = potassium equilibrium potential relative to resting potential; $\bar{g}_{Na}$, V_{Na} are the corresponding quantities for sodium; g_l is the leakage conductance per unit area; and V_l = the equilibrium potential for the leakage current. The coefficients (α's and β's) in the n, m, and h equations are in the standard case

$$\alpha_n(V) = \frac{10 - V}{100(e^{(10-V)/10} - 1)}, \qquad \beta_n(V) = \frac{1}{8} e^{-V/80},$$

$$\alpha_m(V) = \frac{25 - V}{10(e^{(25-V)/10} - 1)}, \qquad \beta_m(V) = 4 e^{-V/18},$$

$$\alpha_h(V) = \frac{7}{100} e^{-V/20}, \qquad \beta_h(V) = \frac{1}{e^{(30-V)/10} + 1}.$$

Choices of interest for the input current density $I(x, t)$ include the following:
(i) Gaussian white noise at a point:

$$\delta(x - x_0)\left(\mu + \sigma \frac{dW(t)}{dt}\right);$$

(ii) Uniform two-parameter Gaussian white noise:

$$\mu + \sigma \frac{\partial^2 W(x, t)}{\partial x \, \partial t};$$

(iii) Poisson excitation and Poisson inhibition at discrete space points:

$$\delta(x - x_E)a_E \frac{dN_E(t)}{dt} - \delta(x - x_I)a_I \frac{dN_I(t)}{dt};$$

(iv) Uniform two-parameter Poisson input:

$$a_E \frac{\partial^2 N_E(x, t)}{\partial x \, \partial t} - a_I \frac{\partial^2 N_I(x, t)}{\partial x \, \partial t},$$

where an explanation of all these symbols has been given above. Naturally generalizations to arbitrary input amplitudes at any point, multiple inputs, and time-dependent inputs can be easily written down. With any of these choices, $V(x, t)$ is a two-parameter random field, as are the activation and inactivation variables $m(x, t)$, $n(x, t)$, and $h(x, t)$.

As is well known, traveling wave solutions of the deterministic HH equations exist and mimic the action potentials. What will solutions be like in the stochastic case? Clearly, solitary waves of V cannot then exist because there is always a chance that V, being a random process, will change shape as it propagates. It may, in fact, randomly decay away to a subthreshold level and randomly grow back again. Naturally, the noise term would have to be very strong for this to happen with a nonnegligible probability. We do expect that the moments of V, m, n, and h will display traveling solitary wave behavior. Also, the probability distributions of V, m, n, and h might also consist of traveling solitary waves called *probalons* or *stochastons*. With such wave phenomena occurring naturally, there is no need to artificially prescribe threshold conditions. Before proceeding with our analysis, we consider a system designed to mimic the HH system.

Stochastic Fitzhugh–Nagumo equations. Because the Hodgkin–Huxley equations are difficult to analyze, a simpler system with only two components has been employed. This system can be written

$$\frac{\partial V}{\partial t} = \frac{\partial^2 V}{\partial x^2} + f(V) - W + I(x, t), \qquad \frac{\partial W}{\partial t} = b(V - \gamma W)$$

where $V = V(x, t)$, $W = W(x, t)$, and f is given by the cubic

$$f(V) = V(1 - V)(V - \tilde{a}).$$

Here $\tilde{a}$ is a constant satisfying $0 < \tilde{a} < 1$ so that f has zeros at 0, $\tilde{a}$, and 1. Roughly speaking, the variables V and m of the Hodgkin–Huxley system are mimicked by the single variable V in the Fitzhugh–Nagumo equations; and the variables n and h are mimicked by the single variable W, which may be called a "recovery variable." The Fitzhugh–Nagumo system, with the appropriate initial and/or stimulus conditions, supports traveling solitary wave solutions, which are supposed to represent action potentials. Random stimulus terms $I(x, t)$, as in the HH case above, lead to a two-dimensional system of random partial differential equations.

6.6.1. Space-clamped systems. Neurophysiologists often try to eliminate diffusive effects by employing what they call a "space-clamp." The word "clamp" to them means "to hold constant," so "space-clamp" means "to hold constant in space." If a quantity is constant in space, then its space derivatives are zero. Thus if we space-clamp a patch of nerve membrane, $V(x, t)$ can be replaced by $V(t)$ and $\partial V/\partial x$ becomes zero.

Hodgkin–Huxley system. In order to obtain a more conventional notation, let us put X for n, Y for m, and Z for h. We are now dealing with the vector $(V(t), X(t), Y(t), Z(t))$ of random processes with $t \geq 0$. Let us also make the substitutions

$$C_1 = \bar{g}_K/C_m,$$
$$C_2 = \bar{g}_{Na}/C_m,$$
$$C_3 = g_l/C_m,$$

and rewrite V_k, V_{Na}, and V_l as v_K, v_{Na}, and v_l, respectively. Putting

(6.16) $$\frac{I}{C_m} = \mu + \sigma \frac{dW_1}{dt}$$

so that the voltage equation is driven by additive Gaussian white noise, the system of equations becomes that of a standard four-dimensional temporally homogeneous Markov process (multidimensional diffusion).

Define the transition probability function

$$P(v, x, y, z, t; \bar{v}, \bar{x}, \bar{y}, \bar{z}, \bar{t}) = \Pr\{V(t) \leq v, X(t) \leq x, Y(t) \leq y,$$
$$Z(t) \leq z \mid V(\bar{t}) = \bar{v}, X(\bar{t}) = \bar{x}, Y(\bar{t}) = \bar{y}, Z(\bar{t}) = \bar{z}\}$$

with corresponding transition density $p(v, x, y, z, t; \bar{v}, \bar{x}, \bar{y}, \bar{z}, \bar{t})$. Then from Chapter 4 we have the following. The transition probability density function p for the space-clamped stochastic Hodgkin–Huxley equations with noise term (6.16) satisfies the backward Kolmogorov equation

$$-\frac{\partial p}{\partial t} = \frac{1}{2}\sigma^2 \frac{\partial^2 p}{\partial \bar{v}^2} + \{c_1(v_K - \bar{v})\bar{x}^4 + c_2(v_{Na} - \bar{v})\bar{y}^3\bar{z} + c_3(v_l - \bar{v}) + \mu\}\frac{\partial p}{\partial \bar{v}}$$

$$+ \{\alpha_n(\bar{v})(1 - \bar{x}) - \beta_n(\bar{v})\bar{x}\}\frac{\partial p}{\partial \bar{x}} + \{\alpha_m(\bar{v})(1 - \bar{y}) - \beta_m(\bar{v})\bar{y}\}\frac{\partial p}{\partial \bar{y}}$$

$$+ \{\alpha_h(\bar{v})(1 - \bar{z}) - \beta_h(\bar{v})\bar{z}\}\frac{\partial p}{\partial \bar{z}},$$

and the forward Kolmogorov equation

$$\frac{\partial p}{\partial t} = \frac{1}{2}\sigma^2 \frac{\partial^2 p}{\partial v^2} - \frac{\partial}{\partial v}(p\{c_1(v_K - v)x^4 + c_2(v_{Na} - v)y^3z + c_3(v_l - v) + \mu\})$$

$$- \frac{\partial}{\partial x}(p\{\alpha_n(v)(1 - x) - \beta_n(v)x\}) - \frac{\partial}{\partial y}(p\{\alpha_m(v)(1 - y) - \beta_m(v)y\})$$

$$- \frac{\partial}{\partial z}(p\{\alpha_h(v)(1 - z) - \beta_h(v)z\}).$$

If the Gaussian white noise current is replaced with a compound Poisson noise

$$\frac{I}{C_m} = \frac{d}{dt}\int_\mathbb{R} u v(t, du),$$

where $v(.,.)$ is a Poisson random measure with

$$E(v(t, du)) = t\Pi(du),$$

then the term $\frac{1}{2}\sigma^2 \partial^2 p/\partial \bar{v}^2$ in the backward equation is replaced by

$$\int_\mathbb{R} p(v, x, y, z, t; \bar{v} + u, \bar{x}, \bar{y}, \bar{z}, \bar{t})\Pi(du) - \Lambda p,$$

where

$$\Lambda = \int_\mathbb{R} \Pi(du).$$

In particular, if there are jumps up of magnitude a_E with intensity λ_E and jumps down of magnitude a_I with intensity λ_I, the term $\frac{1}{2}\sigma^2\partial^2 p/\partial \bar{v}^2$ is replaced by

$$\lambda_E p(v, x, y, z, t; \bar{v} + a_E, \bar{x}, \bar{y}, \bar{z}, \bar{t})$$
$$+ \lambda_I p(v, x, y, z, t; \bar{v} - a_I, \bar{x}, \bar{y}, \bar{z}, \bar{t}) - (\lambda_E + \lambda_I) p.$$

Similarly, the term $\frac{1}{2}\sigma^2 \partial^2 p/\partial v^2$ in the forward equation is replaced by

$$\int_\mathbb{R} p(v - u, x, y, z, t; \bar{v}, \bar{x}, \bar{y}, \bar{z}, \bar{t})\Pi(du) - \Lambda p,$$

in the compound Poisson case, and in particular by

$$\lambda_E p(v - a_E, x, y, z, t; \bar{v}, \bar{x}, \bar{y}, \bar{z}, \bar{t}) + \lambda_I p(v + a_I, x, y, z, t; \bar{v}, \bar{x}, \bar{y}, \bar{z}, \bar{t}) - \Lambda p;$$

then the jumps are $+a_E$ and $-a_I$, with intensities λ_E and λ_I.

We note the following. First, an experiment corresponding to the space-clamped HH system of equations has been performed by Guttman, Feldman, and Lecar (1974), who applied white noise current to a squid axon. Hence, a comparison of theory and experiment is feasible. White noise has also been used as stimulus for nerve cells with more complicated geometries (see, e.g., Bryant and Segundo (1976), Moore and Christensen (1985)). Second, whereas the system of stochastic differential equations is nonlinear, the Kolmogorov equations are linear and therefore amenable to numerical solution by known methods. Furthermore, there is no need to define a threshold function for the nonlinear stochastic system and study first passage times to it. One expects that solutions of the Kolmogorov equations will have probability mass concentrated, in the small σ case, near the deterministic action potential trajectories. For certain values of μ and σ, one might find that the Kolmogorov equations have periodic solutions corresponding to trains of action potentials. Such interesting possibilities await investigation.

Fitzhugh–Nagumo system. For the stochastic space-clamped Fitzhugh–Nagumo equations with additive Gaussian white noise, we put

$$dX = (f(X) - Y + \mu)\,dt + \sigma\,dW, \qquad dY = b(X - \gamma Y)\,dt,$$

where $\{W(t), t \geq 0\}$ is a standard Wiener process and μ and σ are constants. Defining the transition probability function

$$P(x, y, t : \bar{x}, \bar{y}, \bar{t}) = \Pr\{X(t) \leq x,\ Y(t) \leq y \mid X(\bar{t}) = \bar{x},\ Y(\bar{t}) = \bar{y}\},$$

with corresponding density $p(x, y, t; \bar{x}, \bar{y}, \bar{t})$, we have the following.

The transition probability density function for the space-clamped Fitzhugh–Nagumo equations with additive Gaussian white noise satisfies the backward Kolmogorov equation

$$-\frac{\partial p}{\partial \bar{t}} = \frac{1}{2}\sigma^2 \frac{\partial^2 p}{\partial \bar{x}^2} + (f(\bar{x}) - \bar{y} + \mu)\frac{\partial p}{\partial \bar{x}} + b(\bar{x} - \gamma\bar{y})\frac{\partial p}{\partial \bar{y}}$$

and the forward Kolmogorov equation

$$\frac{\partial p}{\partial t} = \frac{1}{2}\sigma^2 \frac{\partial^2 p}{\partial x^2} - \frac{\partial}{\partial x}(p\{f(x) - y + \mu\}) - \frac{\partial}{\partial y}(p\{b(x - \gamma y)\}).$$

The transition density $p(x, y, t; \bar{x}, \bar{y}, \bar{t})$ for the stochastic space-clamped Fitzhugh–Nagumo system with Poisson inputs,

$$dX = (f(X) - Y)\,dt + \int_{\mathbb{R}} u v(dt, du),$$

$$dY = b(X - \gamma Y)\,dt,$$

satisfies the backward Kolmogorov equation

$$-\frac{\partial p}{\partial \bar{t}} = \int_{\mathbb{R}} p(x, y, t; \bar{x} + u, \bar{y}, \bar{t})\Pi(du) - \Lambda p + (f(\bar{x}) - \bar{y})\frac{\partial p}{\partial \bar{x}} + b(\bar{x} - \bar{y})\frac{\partial p}{\partial \bar{y}},$$

where Λ is given above. In particular, if

$$\int_{\mathbb{R}} u v(dt, du) = a_E dN_E - a_I dN_I,$$

then

$$-\frac{\partial p}{\partial t} = \lambda_E p(x, y, t; \bar{x} + a_E, \bar{y}, \bar{t}) + \lambda_I p(x, y, t; \bar{x} - a_I, \bar{y}, \bar{t})$$

$$- (\lambda_E + \lambda_I) p + (f(\bar{x}) - \bar{y})\frac{\partial p}{\partial \bar{x}} + b(\bar{x} - \gamma\bar{y})\frac{\partial p}{\partial \bar{y}}.$$

Furthermore, the corresponding forward Kolmogorov equations are

$$\frac{\partial p}{\partial t} = \int_{\mathbb{R}} p(x - u, y, t; \bar{x}, \bar{y}, \bar{t})\Pi(du) - \Lambda p$$

$$- \frac{\partial}{\partial x}(p\{f(x) - y\}) - \frac{\partial}{\partial y}(p\{b(x - \gamma y)\})$$

and

$$\frac{\partial p}{\partial t} = \lambda_E p(x - a_E, y, t; \bar{x}, \bar{y}, \bar{t}) + \lambda_I p(x + a_I, y, t; \bar{x}, \bar{y}, \bar{t})$$
$$- (\lambda_E + \lambda_I)p - \frac{\partial}{\partial x}(p\{f(x) - y\}) - \frac{\partial}{\partial y}(p\{b(x - \gamma y)\}),$$

respectively.

Note that if any one of these stochastic nonlinear systems has both Poisson and Gaussian white noise, then the corresponding terms in the Kolmogorov equations are simply added, so long as the separate noise sources are independent. For example, if the space-clamped Fitzhugh–Nagumo system is driven by Gaussian and compound Poisson noise, so that we have

$$dX = (f(X) - Y + \mu)\, dt + \sigma\, dW + \int_{\mathbb{R}} uv(dt, du),$$
$$dY = b(X - \gamma Y)\, dt,$$

then the transition density satisfies the forward Kolmogorov equation

$$\frac{\partial p}{\partial t} = \frac{1}{2}\sigma^2 \frac{\partial^2 p}{\partial x^2} - \frac{\partial}{\partial x}(p\{f(x) - y + \mu\}) - \frac{\partial}{\partial y}(p\{b(x - \gamma y)\})$$
$$+ \int_{\mathbb{R}} p(x - u, y, t; \bar{x}, \bar{y}, \bar{t})\Pi(du) - \Lambda p.$$

Stochastic versions of the Frankenhaeuser–Huxley equations, which were designed for a space-clamped situation, are discussed in Tuckwell (1986b).

6.6.2. Perturbative analysis. We will illustrate that some progress can be made with solutions of nonlinear SPDEs involving additive noise, provided the noise term involves a small parameter. The work to be reported in this section has been the subject of some recent articles (Tuckwell (1987a), (1988c)). We show that statistical properties of solutions can be found to any desired accuracy in both scalar and vector reaction-diffusion systems.

Scalar equations. Consider a general nonlinear random reaction-diffusion equation

(6.17) $$U_t = U_{xx} + f(U) + \epsilon(\alpha + \beta W_{xt})$$

in connection with some initial and boundary conditions. In (6.17) $f: R \to R$ is a suitable nonlinear function, ϵ is a small parameter, α and β are constants, and W_{xt} is standard two-parameter white noise. Assume that $U_t = f(U)$ has a critical point at U_0 which is asymptotically stable, i.e., $f(U_0) = 0$, $f'(U_0) < 0$. We put

$$U = U_0 + \sum_{k=1}^{\infty} \epsilon^k U_k,$$

which on substitution in (6.17) gives a sequence of linear SPDEs. We illustrate

with the reduced Fitzhugh–Nagumo equation for which
$$f(U) = U(U-a)(1-U),$$
and set $U_0 = 0$. Then the recursive system is

(6.18) $$U_{1,t} = U_{1,xx} - aU_1 + \alpha + \beta W_{xt},$$

(6.19) $$U_{2,t} = U_{2,xx} - aU_2 + (1+a)U_1^2,$$

(6.20) $$U_{3,t} = U_{3,xx} - aU_3 + 2(1+a)U_1 U_2 - U_1^3.$$
$$\vdots$$

A simplifying feature is that not only are all these equations linear, but they all involve the same Green's function.

Recall §6.5, especially with Neumann conditions at $x = 0$ and $x = L$, which we assume here. U_1 is thus given as

$$U_1(x, t) = \int_0^t \int_0^L G(x, y; t-s)[\alpha \, dy \, ds + \beta \, dW(y, s)].$$

With U_1 at one's disposal, one may readily find U_2 from

$$U_2(x, t) = (1+a) \int_0^t \int_0^L G(x, y; t-s) U_1^2(y, s) \, dy \, ds.$$

The mean of U_2 can be found from the variance of $U_1(x, t)$, which can be easily calculated. To order ϵ^2 we therefore may obtain $E(U(x, t))$ for all x and all t:

$$E(U(x,t)) = \frac{\epsilon \alpha}{a}(1-e^{at}) + \epsilon^2 \bigg[\frac{2(1+a)\alpha^2 e^{-at}}{a^2}\bigg\{\frac{\sinh(at)}{a} - t\bigg\}$$
$$+ \frac{(1+a)\beta^2}{2L}\bigg\{\frac{1}{a^2}(e^{-2at} + 1 - 2e^{-at})$$
$$+ \sum_{n=1}^{\infty} \frac{1}{a+\mu_n^*}\bigg(\frac{1-e^{-at}}{a} + \frac{e^{-2t(a+\mu_n^*)} - e^{-at}}{a+2\mu_n^*}\bigg)$$
$$+ \sum_{n=1}^{\infty} \frac{\phi_{2n}(x)}{a+\mu_n^*}\bigg(\frac{1-e^{-t(a+\mu_{2n}^*)}}{a+\mu_{2n}^*} + \frac{e^{-2t(a+\mu_n^*)} - e^{-t(a+\mu_{2n}^*)}}{a+2\mu_n^* - \mu_{2n}^*}\bigg)\bigg\}\bigg]$$
$$+ O(\epsilon^3),$$

where $\mu_n^* = n^2\pi^2/L^2$ and the ϕ_n's are as before.

The covariance of U at two space-time points can also be evaluated to order ϵ^3 and, hence, the variance along with asymptotic spectral densities (Tuckwell (1987a)). Interestingly, similar equations have been studied in theoretical physics in connection with quantum field theory (Parisi and Wu (1981), Floratos and Iliopoulos (1983), Doering (1985), Jona-Lasinio and Mitter (1985)), and quantum mechanical tunneling (Faris and Jona-Lasinio (1982)). Marcus (1974) has obtained results for equations of the form

$$\dot{u} = \mathcal{L}u + f(u) + n(t),$$

where u takes values in a Hilbert space H, $\mathscr{L}: H \to H$ is a linear time-independent operator, $f: H \to H$ is Lipschitz-continuous, and n is a white noise in H. Results on the convergence of perturbative expansions were included.

Vector case. The Hodgkin–Huxley equations or the Fitzhugh–Nagumo equations with small additive noise take the form

(6.21) $$\mathbf{U}_t = \mathbf{D}\mathbf{U}_{xx} + \mathbf{g}(\mathbf{U}) + \epsilon \mathbf{f}(x, t),$$

where $\mathbf{U}(x, t)$ is an n-dimensional vector, $\mathbf{D}$ is an $n \times n$ constant diagonal matrix of diffusion coefficients, $\mathbf{g}: \mathbb{R}^n \to \mathbb{R}^n$ is a nonlinear function, $\mathbf{f}(x, t)$ is random forcing term, and ϵ is a small parameter. Again, suitable boundary values are assumed to be given. We try a solution of the form

(6.22) $$\mathbf{U} = \mathbf{U}_0 + \sum_{k=1}^{\infty} \epsilon^k \mathbf{U}_k,$$

where $\mathbf{g}(\mathbf{U}_0) = \mathbf{0}$ and $\mathbf{U}_0$ is assumed to be asymptotically stable for $\mathbf{U}_t = \mathbf{g}(\mathbf{U})$. Substituting (6.22) in (6.21) gives a recursive sequence of *linear* systems of SPDEs whose solutions may all be written down using Green's function matrices (Tuckwell (1988d)).

To illustrate, consider the Fitzhugh–Nagumo system with diffusion in both components and additive two-parameter white noise driving the first component:

$$U_t = U_{xx} + U(1 - U)(U - a) - V + \epsilon(\alpha + \beta W_{xt}),$$
$$V_t = V_{xx} + b(U - \gamma V),$$

where $0 < x < L < \infty$, $t > 0$, and the extra diffusion term merely facilitates finding an explicit formula for the Green's function matrix.

With $\mathbf{U}_0 = \mathbf{0}$ we put

$$\mathbf{A} = \begin{pmatrix} -a & -1 \\ b & -b\gamma \end{pmatrix}$$

to get

$$\mathbf{U}_{1,t} = \mathbf{U}_{1,xx} + \mathbf{A}\mathbf{U}_1 + \begin{pmatrix} \alpha + \beta W_{xt} \\ 0 \end{pmatrix},$$

$$\mathbf{U}_{2,t} = \mathbf{U}_{2,xx} + \mathbf{A}\mathbf{U}_2 + \begin{pmatrix} (1+a)U_1^2 \\ 0 \end{pmatrix},$$

$$\mathbf{U}_{3,t} = \mathbf{U}_{3,xx} + \mathbf{A}\mathbf{U}_3 + \begin{pmatrix} 2(1+a)U_1 U_2 \\ 0 \end{pmatrix},$$
$$\vdots$$

where $\mathbf{U}_k = (U_k, V_k)^T$. Letting the Green's function matrix for $\mathbf{u}_t = \mathbf{u}_{xx} + \mathbf{A}\mathbf{u}$ be

$$\mathbf{G}(x, y; t) = e^{\mathbf{A}t} G(x, y; t),$$

where G is the Green's function for the scalar heat equation $u_t = u_{xx}$, we obtain

$$U_1(x, t) = \int_0^L \int_0^t G(x, y; t-s) f_1(y, s) \, ds \, dy,$$

which may be used to find U_2 from

$$U_2(x, t) = \int_0^L \int_0^t e^{A(t-s)} G(x, y; t-s) \binom{(1+a)U_1^2(y, s)}{0} ds \, dy,$$

and so forth. The mean of U has been found to order ϵ^2, and the covariance to order ϵ^3 (Tuckwell (1987b)) as well as the spectral density. The same techniques may be applied to the Hodgkin–Huxley system, but it is more difficult to find G.

CHAPTER 7

The Statistical Analysis of Stochastic Neural Activity

The rise and fall of the depolarization of a nerve cell during an action potential is usually such a rapid event that the sequence of spikes can be regarded as generating a point process. This chapter discusses the quantitative analysis of such point processes. The objects of the experimental studies in this connection come under three main categories:

(a) spontaneous activity, including comparative studies;
(b) stimulus-response relations, including poststimulus time histograms (PSTH) and receptive fields;
(c) cross-correlation in multiple unit recording.

One motivation for these studies, apart from the obvious fundamental relations to theories of perception and behavior, comes from attempts to elucidate the different functional roles of various neurons. The spontaneous activity of a cell characterizes its resting or ground state. Analysis of patterns of spontaneous activity may lead via statistical inference, possibly in conjunction with models such as those considered in the previous four chapters, to estimates of physiological and anatomical properties of nerve cells not obtainable by other means. White noise currents also may be injected into cells in the hope of obtaining physiological and other parameters. Furthermore, it is important to monitor significant changes from time to time in the activity of a cell. Some experiments have focused on the relation between a cell's firing pattern and the state of arousal of the animal. Others have concentrated on the effects of learning, conditioning, drugs, etc. Another broad area of interest is information processing. How does the nervous system detect and identify stimuli? What is its version of statistical hypothesis testing? Unfortunately, we do not have space to address all these issues in depth; we give a cursory summary and direct the reader to relevant references.

7.1. Definition and the basic renewal model.

The simplest assumption is that a sequence of ISIs is a set of independent identically distributed nonnegative random variables $\{T_k, k = 1, 2, \cdots\}$, general symbol T. This generates a renewal process, the theory of which is well known (see, e.g., Cox (1962), Cox and Miller (1965)).

In the present context, we let the ISI distribution function be $\{F(t), t \geq 0\}$ and assume F has a density f. The *spike rate function* $s(t)$ is equivalent to the failure rate function or hazard function in reliability theory. We have

$$s(t) = \frac{f(t)}{1 - F(t)}.$$

The *spike density* is

$$u(t) = \lim_{\Delta t \to 0} \frac{\Pr\{\text{a spike in } (t, t + \Delta t]\}}{\Delta t},$$

which is called the *renewal density* in other contexts. The following are standard relations:

$$u(t) = \frac{d}{dt} E[N(t)] = \sum_{k=1}^{\infty} f_k(t),$$

where $N(t)$ is the number of spikes in $(0, t]$ and f_k is the density of the sum of k intervals. ($u(t)$ is called *expectation density* by some.)

The model we have described assumes a spike at $t = 0$, and this gives an *ordinary renewal process*. If the observation period does not begin at a spike time, the time to the first spike has a different distribution from the intervals $T_2, T_3, \cdots$. The process thus obtained is called a *modified renewal process*, and we let the density of T_1 be f_1 for it. A special case of the modified process is called an *equilibrium renewal process* and for this $f_1(t) = (1 - F(t))/\mu$, where $\mu = E(T)$. This choice for f_1 coincides with the asymptotic $(t \to \infty)$ density of the waiting time for a spike from an arbitrary time point in an ordinary or modified renewal process and is appropriate if a neuron has been spiking for a long time when the observation period (randomly) begins.

It is pointed out that if successive ISIs are independent, then $u(t)$ can be obtained from $f(t)$ since $f_k(t)$ is a $(k-1)$-fold convolution of f with itself. For example, if the waiting time between inputs is Erlangian (λ, m) and one input always leads to a spike in the postsynaptic cell, then

$$u(t) = \lambda e^{-\lambda t} \sum_{k=1}^{\infty} \frac{(\lambda t)^{km-1}}{(km - 1)!}.$$

7.2. The analysis of ISI distributions.

We consider three aspects of the distribution of time intervals between spikes, mainly, but not entirely, based on the hypothesis of a renewal model. The methodology coincides to some extent with that of the analysis of lifetime or survival data (see, e.g., Lawless (1982)).

7.2.1. Estimation of parameters. Assuming that $\{T_1, \cdots, T_n\}$ is a random sample for T, one may obtain values for the sample mean $\bar{T}$ and the sample variance S_T^2. If one suspects that a given parameterized ISI distribution fits the data, then the parameters may be estimated by various methods. *Least-squares methods* are usually implemented with the aid of a computer. The *method of moments* and *maximum likelihood* estimation sometimes yield analytical results, and those that are known are listed in Table 7.1. A missing entry means that a closed form expression was not obtained and uppercase letters signify random variables. Examples of the application of the estimates to actual spike train data are summarized in Tuckwell (1988b, Chap. 10).

TABLE 7.1

ISI Density	Method of Moments	Maximum Likelihood
Exponential, $f(t) = \lambda e^{-\lambda t}, \quad t > 0$	$\hat{\Lambda} = 1/\bar{T}$	same
Exponential with refractory period, $f(t) = \begin{cases} 0, & 0 < t < t_R, \\ \lambda e^{-\lambda(t - t_R)}, & t > t_R \end{cases}$	$\hat{T}_R = \bar{T} - S_T,$ $\hat{\Lambda} = 1/S_T$	$\hat{T}_R = T_1 \wedge T_2 \wedge \cdots \wedge T_n$ $\hat{\Lambda} = \left[\dfrac{\sum (T_i - T_R)}{n}\right]^{-1}$
Exponential with binned data N_i observations in the ith bin, $(i-1)\Delta t < T < i\Delta t, \quad i = 1, \cdots, m$	$\hat{\Lambda} = \dfrac{1}{\Delta t} \ln\left(1 + \dfrac{n_T}{\sum N_i(i - 1)}\right)$ where $n_T = \sum N_i$	—
Gamma density, $f(t) = \dfrac{\lambda(\lambda t)^{m-1} e^{-\lambda t}}{(m-1)!}, \quad t > 0$	$\hat{\Lambda} = \bar{T}/S_T^2,$ $\hat{M} = \bar{T}^2/S_T^2$	—, but see Lawless (1982)
Normal density $f(t) = \dfrac{1}{\sqrt{2\pi\sigma^2}} \exp\left\{\dfrac{-(t - \mu)^2}{2\sigma^2}\right\},$ $t \in \mathbb{R}$	$\hat{\mu} = \bar{T},$ $\hat{\Sigma}^2 = S_T^2$	same
Lognormal $f(t) = \dfrac{1}{t\sqrt{2\pi\sigma^2}} \exp\left\{\dfrac{-(\ln t - \mu)^2}{2\sigma^2}\right\},$ $t > 0$	Convert data to a histogram for $\ln T$ and use estimates for normal.	same
Inverse Gaussian $f(t) = \dfrac{\alpha}{\sqrt{2\pi\beta^2 t^3}} \exp\left\{\dfrac{-(\alpha - t)^2}{2\beta^2 t}\right\},$ $t > 0, \quad E(T) = \alpha, \quad \text{Var}(T) = \alpha\beta^2$	$\hat{\alpha} = \bar{T},$ $\hat{\beta} = S_T^2/\bar{T}$	$\hat{\alpha} = \bar{T},$ $\hat{\beta} = \left[\bar{T}\left(-1 + \dfrac{\bar{T}}{n} \sum_{k=1}^{n} \dfrac{1}{T_k}\right)\right]$

Unfortunately, there are no known theoretical ISI densities for the more realistic neuronal models which incorporate exponential decay or cable properties. Nevertheless, it was possible to apply the method of moments to data on cat cochlear nucleus cells to obtain parameters in Stein's model with excitation only (Tuckwell and Richter (1978)). The inverse Gaussian density has been fitted to ISI data by Gerstein and Mandelbrot (1964), Nilsson (1977), and Correia and Landolt (1979). However, it does not seem worthwhile to persist in using this density simply because analytical formulas are available for it and not for more realistic models.

7.2.2. Testing and comparing distributions. In the analysis of neural data, statistical tests may be carried out to test whether a given distribution provides a good fit to some data or to compare two or more empirical distributions.

(i) *Goodness-of-fit test to a given distribution.* Among the various possibilities of tests are the nonparametric χ^2-*test* and the *Kolmogorov–Smirnov test*. To implement the latter, an empirical distribution function is computed:

$$\tilde{F}_n(t) = \frac{\text{number of } T_k\text{'s} \leq t}{n},$$

and compared with the hypothesized distribution function. The distribution-free test statistic employed is

$$D_n = \sup_{t>0} |\tilde{F}_n(t) - F(t)|.$$

Note that superior tests are available for testing against certain particular kinds of distribution functions (Lawless (1982, Chap. 9)).

(ii) *Detecting changes in an* ISI *distribution.* In many situations, a comparison is sought between (a) the firing pattern of a cell at one time or under a given set of circumstances, and (b) the firing pattern at another time or under different conditions. For example, to compare two empirical distribution functions $\tilde{F}_{n_1}$ and $\tilde{F}_{n_2}$, we employ the *two-sample Kolmogorov-Smirnov statistic*:

$$D_{n_1, n_2} = \sup_{t>0} |\tilde{F}_{n_1}(t) - \tilde{F}_{n_2}(t)|.$$

Critical values of D_{n_1,n_2} are given in Pearson and Hartley (1972). A test that may be used for comparing several data sets is the nonparametric *Kruskal–Wallis test* (called Wilcoxon test for two data sets). If the ISIs are approximately normal, we may use a *two-sample t-test* to compare means. All the above tests have been applied to neural spike train data.

7.2.3. The classification and interpretation of ISI distributions. On the examination of about 200 ISI histograms from various neurons in various animals, it has been possible to distinguish 10 different types of distribution. These are labeled 1–7, 8A, 8B, 8C, 9, and 10; they are illustrated in Fig. 7.1.

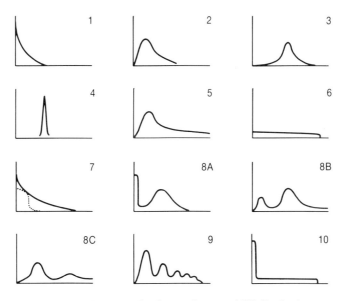

FIG. 7.1. *Illustrating the classes of neuronal ISI distributions.*

Unimodal densities, 1–6.

1. *Exponential.* This is frequently encountered and may signify Poisson excitation with only one EPSP required to elicit an action potential.

2. *Low-order gamma-like.* This is the most common ISI distribution. Among the various possibilities are (a) the waiting time between inputs has this density and every input causes a spike (this applies of course to all 10 classes); and (b) a few EPSPs are required to reach threshold. It is unlikely that the cell is receiving significant amounts of inhibition.

3. *Normal.* There is an overlap with type 2 (and type 4), but here there is a definite lack of very short and very long intervals. It is likely that a large number of EPSPs are required to make the cell fire and that there is little inhibition.

4. *Delta-like.* This is similar to a point mass or a Gaussian with a very small standard deviation. The cell is firing very regularly, as do pacemaker cells. An intrinsic mechanism may be involved, or there may be a periodic external stimulus.

5. *Long-tailed.* These may resemble types, 1–3 at small to moderate ISIs, but the density has a very long tail at large times. A possible reason for the long tail is the presence of large quantities of inhibition.

6. *Uniform.* The ISI density is practically flat with no preferred intervals less than some upper limit. A possible explanation is that the cell operates in several modes so that the ISI density is a linear combination of several underlying densities with a varying number of input events capable of producing an action potential (cf. the spike density for an exponentially distributed ISI).

Bimodal and multimodal, 7–10.

7. *Superposition of two exponentials.* This is not really bimodal but is a linear combination of two different exponential densities. A model that predicts such a distribution was given by Smith and Smith (1965). Spikes occur as a Poisson process with intensity v, but this is blocked for periods which are exponentially distributed with mean $1/\mu$. The periods of blocking are separated by time intervals which are exponentially distributed with mean $1/\lambda$. A spike always accompanies a transition from the blocked to unblocked states, but not on the reverse transitions. It is not hard to show that the ISI density is then

$$f(t) = ve^{-(\lambda+v)t} + \frac{\lambda\mu}{v+\lambda-\mu}(e^{-\mu t} - e^{-(\lambda+v)t}), \quad t > 0.$$

Smith and Smith (1965) reported 40 cells in the *isolated* cerebral cortex of cat which had this kind of ISI distribution. Note that the coefficient of variation is greater than one.

8. *Bimodal.* Types 8A, 8B, and 8C are bona fide bimodal densities. If the time of a second peak is twice the first, then there is a random deletion of spikes, which may or may not be due to recording techniques. Again, a cell may have two different modes of spiking. Otherwise, a peak at small times may indicate bursting, i.e., doublets, triplets, or short trains of high frequency spikes. Some CNS cells may burst at frequencies of up to 1,200/second.

9. *Multimodal.* Multimodality may arise if sometimes one, sometimes two, etc., spikes are missed, as then the ISI density is a linear combination of those for T, $2T$, etc. Such blocking may be due to random inhibition.

10. *L-shaped.* This unusual density combines bursting with the uniform distribution of type 6.

7.3. Tests for temporal structure.

Inter alia tests for temporal structure include those for: (i) stationarity, (ii) independence, (iii) renewal process, and (iv) Poisson process. We will discuss these briefly. Much relevant material is contained in Cox and Lewis (1966) and in a neural context in Yang and Chen (1978). See also Tuckwell (1988b, Chap. 10).

7.3.1. Stationarity. A simple yet crude method for testing for stationarity is to compare the mean ISIs from different parts of the record. A better method is to compare instead the empirical distribution functions as outlined in §7.2.2. Another method popular with neurophysiologists is the *runs test* (see, e.g., Walpole and Myers (1985)). A graphical method is that of *cusums* (Muschaweck and Loevner (1978)). Let $T_1, \cdots, T_n$ be a sequence of ISIs. The cusum at r is the random variable

$$Q_r = r\left(\frac{1}{r}\sum_{k=1}^{r} T_k - \frac{1}{n}\sum_{k=1}^{n} T_k\right).$$

For a stationary process $E(Q_r) = 0$ for all r, but if the mean is μ_1 for the first m intervals and μ_2 for remaining $n - m$, then the slope of Q_r plotted against r changes from $\mu_1 - \mu$ to $\mu_2 - \mu$, where μ is the grand mean

$$\mu = \frac{m}{n}\mu_1 + \left(\frac{n-m}{n}\right)\mu_2.$$

7.3.2. Independence. An easily executed test for independence consists of examining a sequence of *autocorrelation coefficients*

$$\rho_j = \frac{\text{Cov}(T_k, T_{k+j})}{\sigma^2}, \quad j = 1, 2, \cdots.$$

An estimator for ρ_j is

$$R_j = \frac{\sum_{k=1}^{n-j}(T_k - \bar{T})(T_{k+j} - \bar{T})}{\sum_{k=1}^{n}(T_k - \bar{T})^2}.$$

Provided $j \ll n$ and n is large, R_j is, under the assumption of independence, about normally distributed with mean zero and variance $1/n$ (see, e.g., Chatfield (1980)). A plot of r_j versus j, called a *serial correlogram*, is often employed to quickly ascertain whether any of the r_j's are significant.

A test for independence was proposed by Rodieck, Kiang, and Gerstein (1962), which utilized the facts that if T_k and T_{k+1} are independent and identically distributed, then

$$E(T_{k+1} \mid t_1 < T_k \leq t_1 + \Delta t_1) = E(T_{k+1}) = \mu,$$
$$E(T_k \mid t_2 < T_{k+1} \leq t_2 + \Delta t_2) = E(T_k) = \mu.$$

Estimates of these two quantities for various t_1, t_2 are called *column means* and *row means*, respectively, and should be about constant when successive intervals are independent. Yang and Chen (1978) called this a *conditional means test* and construed it as a standard one-way analysis of variance for testing the equality of several means (F test). Finally, we mention *spectral methods* involving, for example, the discrete Fourier transform of $\{\rho_j\}$ which reduces to

$$f^*(\omega) = \frac{1}{2\pi}\left(1 + \sum_{j=1}^{\infty} \rho_j \cos(j\omega)\right), \quad \omega \in [0, \pi].$$

f^* is called the *spectrum of the intervals* and if $\rho_j = 0$, observed values of f^* will be approximately flat. See Cox and Lewis (1966) and Kwaadsteniet (1982).

7.3.3. Renewal process. If $\{N(t), t \geq 0\}$ counts the number of spikes in $(0, t]$ and the renewal model is valid, then the asymptotic normality of $N(t)$ as $t \to \infty$ may be used to test for a renewal process (Yang and Chen (1978)). In one study, 64 percent of cells studied satisfied the renewal hypothesis (Correia and Landolt (1979)), whereas the figure was 70 percent in another study (Ekholm and Hyvarinen (1970)).

7.3.4. Poisson process. Tests for simple homogeneous Poisson processes, important for applications at neuromuscular junctions, (Chapter 2), as well as spike trains, were given in Cox and Lewis (1966). Processes with time-dependent intensity are transformed to standard Poisson processes. The intensity function may be estimated by maximum likelihood (Cox and Hinkley (1975)).

7.4. Parameter estimation for diffusion processes representing nerve membrane potential.

In the event that intracellular recording from a nerve cell is available, it is possible that parameter estimation could be done from sample paths of the membrane potential. In particular, suitable methods have been given when diffusion processes such as Wiener processes or OUPs were adequate (Lánský (1983)). Two procedures have been proposed.

(i) *Maximum likelihood on sample paths.* Feigin (1976) gave a cogent description of this technique which is based on martingale limit theory. For the OUP $dX = (\alpha - \beta X)\,dt + \sigma\,dW$ the estimates of α, β, and σ from sample paths on $[0, t]$ are

$$\hat{\alpha} = \left[B(t) \int_0^t X^2(s)\,ds - \frac{1}{2} C(t) \int_0^t X(s)\,ds \right] \Big/ D(t),$$

$$\hat{\beta} = \left[B(t) \int_0^t X(s)\,ds - \frac{t}{2} C(t) \right] \Big/ D(t),$$

$$\hat{\sigma} = \frac{1}{t} \int_0^t (dX(s))^2,$$

where

$$B(t) = X(t) - X(0),$$
$$C(t) = X^2(t) - X^2(0) - t,$$
$$D(t) = t \int_0^t X^2(s)\,ds - \left(\int_0^t X(s)\,ds \right)^2.$$

(ii) *Maximum likelihood on a transition probability density.* Let the observed values of the depolarization be $X(t_k) = x_k$, $k = 1, 2, \cdots, n$; and assume $X(0) = 0$. For example, using the transition probability density for the OUP $dX = (\alpha - X)\,dt + \sigma\,dW$ (as given in Chapter 5), we find on maximizing the likelihood, the estimates

$$\hat{\alpha} = \left(\sum_{k=1}^n x_k \right) \Big/ \sum_{k=1}^n (1 - e^{-2t_k}),$$

$$\hat{\sigma}^2 = \frac{2}{n} \sum_{k=1}^n \left(\frac{x_k - \hat{\alpha}(1 - e^{-t_k})}{\sqrt{1 - e^{-t_k}}} \right)^2.$$

7.5. Analysis of the simultaneous activities of two or more neurons.

It is often of interest to examine the ongoing activities of several cells. If intracellular recording is possible from many cells at once, then the analysis may proceed directly (e.g., Bryant, Marcos, and Segundo (1973)). Otherwise, there is a possible *separation problem* to determine which spikes came from a given cell (see, e.g., Millechia and McIntyre (1978), Dinning and Sanderson (1981)).

For two spike trains, we define a *joint spike density*

$$u(t_1, t_2) = \lim_{\Delta t_1, \Delta t_2 \to 0^+} \Pr\{\text{a spike of neuron 1 in } (t_1, t_1 + \Delta t_1]$$

and a spike of neuron 2 in $(t_2, t_2 + \Delta t_2]\}/\Delta t_1 \Delta t_2$.

For independent trains, $u(t_1, t_2) = u_1(t_1)u_2(t_2)$. An estimate of $u(t_1, t_2)$ can be made in the usual way, and various hypotheses concerning its form can be tested by using a χ^2 goodness-of-fit test.

Assuming at least second-order stationarity, we define the *cross-intensity function*,

$$u_{12}(t) = \lim_{\Delta t \to 0} \frac{\Pr\{\text{neuron 2 spikes in } (s + t, s + t + \Delta t] \mid \text{neuron 1 spikes at } s\}}{\Delta t}$$

and similarly for $u_{21}(t)$. If the spike trains are stationary so that $u_k(t) = 1/\mu_k$, $k = 1, 2$, where μ_k is a mean interval, then (Perkel, Gerstein, and Moore (1967))

$$\frac{u_{12}(t)}{\mu_1} = \frac{u_{21}(-t)}{\mu_2}.$$

The cross-intensity function (also called cross-correlation function) may be estimated from a *cross-correlation histogram*. Many such histograms have appeared in the literature. Particularly interesting are those obtained in visual receptive field analysis (Toyama and Tanaka (1984)). The use of *cross-periodograms* was advocated in this connection by Jenkins (in Bartlett (1963)). This and other spectral methods have been developed by Brillinger (1975), (1987), (1988). Tests for the independence of two spike trains are given in Perkel, Gerstein, and Moore (1967). Habib and Sen (1985) have discussed methods for analyzing nonstationary spike trains. Methods for analyzing input-output temporal relations using Wiener kernel-type expansions for point processes have been developed by Brillinger (1975) and Krausz (1975). For more detailed references, see Tuckwell (1988b).

Finally, we draw attention to the problems of interpretation and analysis of *poststimulus time histograms* (PSTH), which are obtained in stimulus-response experiments on single neurons. The underlying neural dynamics may be extremely complex, and this is a neglected area. Some studies put PSTHs in the same category as cross-correlation. Sometimes if a target cell is spontaneously active, we have the problem of distinguishing the signal from the noise—a process broadly called *filtering* (see, e.g., Sanderson (1980)).

CHAPTER 8

Channel Noise

The mathematical model of Hodgkin and Huxley for the dynamical properties of nerve membrane potential emphasized the essential role of ion channels. Since 1970, when Katz and Miledi first indicated that noise measurements might yield information about ion channels themselves, the neurophysiological world has been in a state of intense activity in a quest to attain such information. We will take a brief look at some of these developments and direct the reader to many excellent reviews and substantive research articles. Although the mathematics involved is fairly standard, its importance lies in its application rather than in its sophistication.

8.1. Introduction.

Four kinds of noise are distinguished in connection with electrical measurements in biophysics. We will not be concerned with three of these: *thermal noise, shot noise,* and *1/f noise* (see, e.g., Verveen and De Felice (1974) for discussion and references; and Hooge (1976) and Neumcke (1978) for reviews of 1/f noise in particular). Here only channel noise is considered where the variability is due to the apparently random opening and closing of ionic channels. First we will review a few basic facts concerning channels (see, e.g., Hille (1984) for a complete picture).

8.1.1. Ionic channels. An ionic channel is a hole or pore in a cell membrane. It is named for the ionic species to which in usual circumstances it is the most permeable, e.g., K channel, Na channel. In physical structure a channel has been deduced to be a large molecule whose different configurations correspond to the channel's being in closed or open states. A component called a *gate* determines in which state a channel is.

A broad division of ionic channels is into *synaptic* and *nonsynaptic* (or nerve). Synaptic channels are sensitive to neurotransmitters such as acetylcholine, glutamate, etc.; some channels are sensitive to voltage. Nonsynaptic channels are usually sensitive to certain drugs, which is the basis of anesthetics. Channel diameters are believed to be about 6 Å (1 Å = 10^{-10} m), which may be compared with membrane thickness of about 100 Å. Their spatial densities vary; Na channels, for example, occur at about 300 per square micron in squid axon and 3,000 per square micron at frog nodes.

A major breakthrough occurred in 1976, when recordings were made from a single channel (Neher and Sakmann (1976)). The current showed random steps from zero to several picoamps reflecting closed and open states. Some of the quantities sought are the unitary conductance, the characteristics of the open time, and the channel density. Mathematical models are helpful for the following reasons: (a) not all channels can be isolated, so that inference from gross recordings is used to obtain single channel parameters; this is called *fluctuation analysis*; (b) the various reaction schemes for binding with drugs and transitions among various open and closed states lead to different predictions, which we hope can be distinguished experimentally.

8.2. Continuous time Markov chains.

Models for the opening and closing of individual channels take the form of continuous time Markov chains. These have been energetically employed in the present context (Neher and Stevens (1977), Colquhoun and Hawkes (1977), (1981), (1982)). If we let $\{X(t), t \geq 0\}$ be a Markov process with states $\{1, 2, \cdots, n\} = S$, the matrix $\mathbf{P}(t)$ of transition probabilities, assumed stationary,

$$P_{ij}(t) = \Pr\{X(s+t) = j \mid X(s) = i\}, \quad i, j \in S, \quad s, t \geq 0,$$

satisfies the backward

$$\frac{d\mathbf{P}}{dt} = \mathbf{AP}$$

and forward

$$\frac{d\mathbf{P}}{dt} = \mathbf{PA}$$

Kolmogorov equations, where the infinitesimal matrix $\mathbf{A}$ is defined as

$$\mathbf{A} = \lim_{\Delta t \to 0^+} \frac{\mathbf{P}(\Delta t) - \mathbf{I}}{\Delta t},$$

where $\mathbf{I}$ is an identity matrix. The following are standard results:

(i) The time interval spent in any (nonabsorbing) state once entered is exponentially distributed with mean $-a_{ii}^{-1}$.

(ii) If $\mathbf{p}(t)$ is a row vector of absolute probabilities,

$$p_i(t) = \Pr\{X(t) = i\}, \quad i \in S,$$

then

$$\mathbf{p}(t) = \mathbf{p}(0)\mathbf{P}(t).$$

(iii) Since $\dot{\mathbf{p}} = \mathbf{pA}$, if a limiting stationary distribution exists, it can be found by solving $\mathbf{pA} = \mathbf{0}$.

8.2.1. One two-state channel.
The classical theory of drug action consists of the following simple reaction scheme:

$$A + R \underset{\alpha}{\overset{\beta}{\rightleftharpoons}} AR,$$

where A represents an unbound agonist molecule, R represents a receptor corresponding to a closed state (state two) and AR is the agonist-receptor complex representing an open channel configuration (state one). In a deterministic theory, α and β are rate constants, but in the stochastic scheme they are defined through

$$\Pr\{\text{channel open at } t \to \text{closed at } t + \Delta t\} = \alpha \Delta t + o(\Delta t),$$

$$\Pr\{\text{channel closed at } t \to \text{open at } t + \Delta t\} = \beta \Delta t + o(\Delta t).$$

The matrix $\mathbf{A}$ in this case is just

$$\mathbf{A} = \begin{bmatrix} -\alpha & \alpha \\ \beta & -\beta \end{bmatrix},$$

and the probability of being open at t is

$$p_1(t) = \frac{\beta}{\alpha + \beta} + \left(p_1(0) - \frac{\beta}{\alpha + \beta}\right) e^{-(\alpha + \beta)t},$$

whereas the probability of being closed is

$$p_2(t) = \frac{\alpha}{\alpha + \beta} + \left(p_2(0) - \frac{\alpha}{\alpha + \beta}\right) e^{-(\alpha + \beta)t}.$$

If $\{X(t)\}$ is the number of open channels, then $p_1(t) = \Pr\{X(t) = 1\}$ and $p_2(t) = \Pr\{X(t) = 0\}$. As $t \to \infty$, X clearly has a stationary distribution $[\beta \ \alpha]/(\alpha + \beta)$. Alternatively, if $p_1(0)$ and $p_2(0)$ equal the stationary values, then the resulting process in this special case, which we denote by $\hat{X}(t)$, is strictly stationary. If the unitary conductance (open channel) is a, then the conductance at t is just a times $\hat{X}(t)$. It is desirable to carry the experiment out under a *voltage clamp* in order to eliminate capacitative currents and other spurious effects. In this case, if the clamp voltage is V and the Nernst potential for the ionic species under study is V_i, then the current is

$$\hat{I}(t) = a(V - V_i)\hat{X}(t).$$

This current has the following properties:
(a) *Mean.*
$$E(\hat{I}(t)) = \frac{a(V - V_i)\beta}{\alpha + \beta}.$$

(b) *Variance.*
$$\text{Var}(\hat{I}(t)) = \frac{a^2(V - V_i)^2 \alpha \beta}{(\alpha + \beta)^2}.$$

(c) *Covariance.*

$$\mathrm{Cov}(\hat{I}(s), \hat{I}(t)) = \frac{a^2(V - V_i)^2 \alpha \beta_e}{(\alpha + \beta)^2} \exp[-(\alpha + \beta)(t - s)].$$

(d) *Spectral density.*

$$f_I(\omega) = \frac{a^2(V - V_i)^2 \alpha \beta}{\pi(\alpha + \beta)} \cdot \frac{1}{(\alpha + \beta)^2 + \omega^2},$$

which is called a *Lorentzian*. The frequency ω_c satisfying

$$f(\omega_c) = \tfrac{1}{2} f(0)$$

is called the *corner frequency* and here is given by $\omega_c = \alpha + \beta$. Thus $\alpha + \beta$ may be easily obtained from the spectral density, which is routinely obtained experimentally by the use of narrow band filters.

(e) *Waiting times.* The times spent in the open and closed states are exponentially distributed with means $1/\alpha$ and $1/\beta$, respectively. The sum of these (independent) random variables is the time between consecutive openings or closings with density

$$\frac{\alpha \beta}{\beta - \alpha}(e^{-\alpha t} - e^{-\beta t}), \qquad \alpha \neq \beta,$$

and this is also of interest to physiologists.

From the expressions and experimental records of the current through a single channel one may, if the model is valid and $V - V_i$ is known, estimate α and β from the distributions of time spent in the open and closed states, the unitary conductance a from the mean, and then check the value of $\alpha + \beta$ from the spectral density and a from the variance.

8.2.2. One n-state channel, $n \geq 2$. It has been found in some cases that a two-state Markov chain is inadequate. For example, the spectral density cannot be fitted to a single Lorentzian. The above model is then extended to include many states. Such an extension has been elaborated on in the references given at the beginning of this section. One proposed scheme is to have r open states and $n - r$ closed states connected in the following manner:

$$\underbrace{1 \rightleftharpoons 2 \rightleftharpoons 3 \rightleftharpoons \cdots r}_{\text{open}} \rightleftharpoons \underbrace{r + 1 \rightleftharpoons \cdots \rightleftharpoons n - 1 \rightleftharpoons n}_{\text{closed}},$$

where transitions occur only between neighboring states. The infinitesimal matrix of $\{X(t)\}$ can be partitioned thus:

$$\mathbf{A} = \begin{bmatrix} \text{open-open} & \vdots & \text{open-closed} \\ \cdots\cdots\cdots & \vdots & \cdots\cdots\cdots \\ \text{closed-open} & \vdots & \text{closed-closed} \end{bmatrix},$$

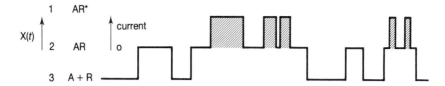

FIG. 8.1. *Showing transitions among the various states in the three-state example. Only the shaded portions are observable as current flow.*

and is in fact tridiagonal. In general, the absolute probabilities can be written

$$\mathbf{p}(t) = \mathbf{p}(\infty) + \mathbf{p}(0) \sum_{k=2}^{n} \mathbf{A}_k e^{\lambda_k t}.$$

Here the matrix $\mathbf{A}_k$ is given by

$$\mathbf{A}_k = \mathbf{u}_k \mathbf{v}_k,$$

where $\mathbf{u}_k$ is a column eigenvector of $\mathbf{A}$ with eigenvalue λ_k, and $\mathbf{v}_k$ is the kth row of the matrix

$$[\mathbf{u}_1 \cdots \mathbf{u}_n]^{-1}.$$

The covariance function of $\{\hat{X}(t)\}$ is of the form

$$\mathrm{Cov}(\hat{X}(s), \hat{X}(t)) = \mathrm{const.} \sum_{k=2}^{n} c_k e^{\lambda_k t},$$

so the spectral density is the sum of $n-1$ Lorentzians.

Example. The following scheme introduced by del Castillo and Katz (1957) was proposed by Magleby and Stevens (1972) for the acetylcholine-activated channel in the end-plate of frog neuromuscular junction:

$$\underset{\substack{\text{receptor}\\ \text{(closed)}\\ 3}}{R} + \underset{\substack{\text{agonist}}}{A} \underset{k_{-1}}{\overset{k_1}{\rightleftharpoons}} \underset{\substack{\text{closed}\\ \text{complex}\\ 2}}{AR} \underset{\alpha}{\overset{\beta}{\rightleftharpoons}} \underset{\substack{\text{open}\\ \text{complex}\\ 1}}{AR^*}.$$

The infinitesimal matrix for the corresponding Markov chain is

$$\mathbf{A} = \begin{bmatrix} -\alpha & \vdots & \alpha & & 0 \\ \cdots & \cdots & \cdots & \cdots & \cdots \\ \beta & \vdots & -(\beta + k_2) & & k_2 \\ \cdots & \vdots & \cdots & \cdots & \cdots \\ 0 & \vdots & k_1 x_A & & -k_1 x_A \end{bmatrix},$$

where x_A is the concentration of the agonist. Under voltage clamp sample paths for the current for this model appear as in Fig. 8.1.

8.3. Many two-state channels.

We now consider a patch of membrane with many channels scattered over it and assume that a recording gives the sum of the currents through the individual channels. The sample paths become relatively smooth, as is illustrated nicely in Lecar and Sachs (1981). Let there be N identical channels which act independently and have only two states, as in §8.2.1. Then the total current is, in the stationary regime,

$$\hat{I}(t) = N\hat{I}_i(t),$$

where each of the $\hat{I}_i$ are as described earlier. The mean, variance, covariance function, and spectral density are now just N times those for a single channel. This simple observation has some powerful consequences, as follow:

(i) The multichannel spectral density has the same form as that for a single channel. Hence, the value of $\alpha + \beta$ may be obtained from the macroscopic result.

(ii) The ratio of the multichannel variance to the mean is the same as for one channel:

$$\frac{\text{Var}(\hat{I}(t))}{E(\hat{I}(t))} = \frac{a(V - V_i)}{\alpha 1 + (\beta/a)} \cong a(V - V_i),$$

where it is assumed that $\alpha \gg \beta$, i.e., the average time spent in the open state is much less than that in the closed state. Thus the value of a, the individual channel conductance, can be estimated from the multichannel recording. Since $\alpha + \beta \cong \alpha$, the value of α may also be estimated.

8.3.1. Birth and death process. The number of open channels in a patch is a birth and death process, which we denote by $\{X_N(t)\}$, as studied in this context by Lam and Lampard (1981); a more complicated model was analyzed by Jackson (1985). For the birth and death process, the infinitesimal transition probabilities are

$$\Pr\{X_N(t + \Delta t) = j \mid X_N(t) = k\} = \begin{cases} 1 - \Delta t[k\alpha + (N - k)\beta] + o(\Delta t), & j = k, \\ (N - k)\beta \Delta t + o(\Delta t), & j = k + 1, \\ k\alpha \Delta t + o(\Delta t), & j = k - 1. \end{cases}$$

Putting

$$p_k(t) = \Pr\{X_N(t) = k \mid X_N(0) = N_0\},$$

where $0 \le k$, $N_0 \le N < \infty$, we find

$$\frac{dp_k}{dt} = (N - k + 1)\beta p_{k-1} - (k\alpha + (N - k)\beta)p_k + (k + 1)\alpha p_{k+1},$$

with $p_{-1} = p_{N+1} = 0$ by definition. The probability-generating function

$$\phi(v, t) = \sum_{k=0}^{N} v^k p_k(t)$$

can be shown to satisfy

$$\phi_t = \beta N(v - 1)\phi + [\alpha + (\beta - \alpha)v - \beta v^2]\phi_v.$$

Solving this equation with initial condition $\phi(v, 0) = v^{N_0}$ gives

$$\phi(v, t) = (\alpha + \beta)^{-N}[\alpha + \beta v + \alpha(v - 1)e^{-(\alpha+\beta)t}]^{N_0}$$
$$\times [\alpha + \beta v - \beta(v - 1)e^{-(\alpha+\beta)t}]^{N-N_0}.$$

From this, an explicit expression for the probability of k open channels at time t may be obtained:

$$p_k(t) = \frac{(\alpha T + \beta)^{N_0}(\beta - \alpha T)^{N-N_0}}{(\alpha + \beta)^N}$$

$$\times \sum_{j=0}^{N} \binom{N_0}{k-j}\binom{N-N_0}{j}\left(\frac{\alpha - \alpha T}{\beta + \alpha T}\right)^{N_0-k+j}\left(\frac{\alpha + \beta T}{\beta - \beta T}\right)^{N-N_0-j},$$

where T is defined as

$$T = e^{-(\alpha+\beta)t}.$$

The mean, variance, covariance, and asymptotic spectral density of this birth and death process can be easily obtained from the single channel results (i.e., for $N = 1$). Furthermore, the birth and death process is asymptotically stationary with a binomial distribution:

$$\Pr\{\hat{X}_N(t) = k\} = \binom{N}{k}\left(\frac{\beta}{\alpha + \beta}\right)^k \left(\frac{\alpha}{\alpha + \beta}\right)^{N-k}, \quad k = 0, 1, \cdots, N.$$

8.3.2. Diffusion approximations. When the number of ionic channels is large, so that the sample paths of $\{X_N(t)\}$ become relatively smooth, we may approximate the birth and death process by a continuous one. To this end we construct two diffusion approximations $\{Y_N(t)\}$ and $\{Z_N(t)\}$ (Tuckwell (1987d)).

Weak convergence-based diffusion. The first diffusion approximation is based on the following result of McNeil and Schach (1973). In the definition of weak convergence, here the uniform metric is employed rather than a Skorohod metric, as in the results of Chapter 5 (see Stone (1963)).

THEOREM 8.1. *Let* $\{\{X_N(t)\}, N = 1, 2, \cdots\}$ *be a sequence of birth and death processes as defined above. Then*

$$\frac{X_N(t) - (\beta N/(\alpha + \beta))}{\sqrt{N}} \xrightarrow{w} \mathcal{U}(t),$$

where $\{\mathcal{U}(t)\}$ *is an Ornstein–Uhlenbeck process with stochastic differential equation*

$$d\mathcal{U} = -(\alpha + \beta)\mathcal{U}\,dt + \sqrt{\frac{2\alpha\beta}{\alpha + \beta}}\,dW.$$

The approximating process $\{Y_N(t)\}$ is obtained by rearranging the above result:

$$Y_N(t) = \frac{\beta N}{\alpha + \beta} + \sqrt{N}\,\mathcal{U}(t).$$

Just as with the birth and death process, Y_N has a stationary version $\{\hat{Y}_N(t)\}$, whose distribution at t is the asymptotic distribution of $\hat{Y}_N$. Thus $\hat{Y}_N(t)$ is a normal random variable with density

$$\tilde{p}_{\hat{Y}_N}(y) = \frac{\alpha+\beta}{\sqrt{2\pi\alpha\beta N}} \exp\left\{-\frac{(y-(\beta N/(\alpha+\beta)))^2(\alpha+\beta)^2}{2\alpha\beta N}\right\},$$

and the same mean and variance as $\hat{X}_N(t)$.

Standard diffusion approximation. The construction of a diffusion process with the same first two infinitesimal moments as the original process gives one whose stochastic differential equation is

$$dZ_N = (\beta N - (\alpha+\beta)Z_N)\,dt + \sqrt{\beta N + (\alpha-\beta)Z_N}\,dW.$$

Thus the infinitesimal mean vanishes at

$$z_{1,N} = \frac{\beta N}{\alpha+\beta},$$

whereas the infinitesimal variance vanishes at

$$z_{2,N} = \frac{-\beta N}{\alpha+\beta}.$$

In Feller's classification of boundary points, $z_{1,N}$ is regular and $z_{2,N}$ is entrance. A suitable choice for the range of Z_N is $(z_{2,N}, \infty)$, since it will be found that the process then spends very little time outside $[0, N]$. The stationary density $\tilde{p}(z)$ of Z_N must satisfy the steady-state Kolmogorov equation

$$\frac{1}{2}\frac{d^2}{dz^2}([\beta N + (\alpha-\beta)z]\tilde{p}) - \frac{d}{dz}([\beta N - (\alpha+\beta)z]\tilde{p}) = 0, \qquad z > z_{2,N}.$$

Integrating twice gives

$$\tilde{p}(z) = k_1\left(\int_0^z e^{Ax}(B+Cx)^{-D}\,dx\right)e^{-Az}(B+Cz)^{D-1} + k_2 e^{-Az}(B+Cz)^{D-1},$$

where

$$A = \frac{2(\alpha+\beta)}{\alpha-\beta},$$

$$B = \beta N,$$

$$C = \alpha - \beta,$$

$$D = \frac{4\alpha B}{(\alpha-\beta)^2}.$$

It is found that $\tilde{p}$ can only be positive for all $z > z_{2,N}$ if $k_1 = 0$. Then k_2 is found by normalizing $\tilde{p}$. This gives

$$\tilde{p}(z) = \frac{A^D C^{1-D} e^{-AB/C}}{\Gamma(D)} e^{-Az}(B+Cz)^{D-1}.$$

It is indeed fortunate that the mean and variance of this density are identical to those of $\hat{X}_N(t)$ and $\hat{Y}_N(t)$. In Tuckwell (1987b) a comparison is made of the stationary distributions of the original birth and death process to the two diffusion approximations.

8.4. Voltage noise.

When a voltage clamp is not employed, as in the original experiment of Katz and Miledi (1972), the electrical potential across a cell membrane fluctuates under the influence of the random openings and closings of ionic channels. If we consider a space-clamped patch with capacitance C_m and nonfluctuating conductance G_m, the voltage $V(t)$ will be determined by the equation

$$C_m \frac{dV}{dt} + G_m V = \sum_{i=1}^{m} (V_i - V) g_{N_i}(t),$$

where it is assumed there are m ionic species with Nernst potentials V_i, $i = 1, \cdots, m$, and $g_{N_i}(t)$ is the conductance due to the N_i channels of the ith species. If the unitary conductances are a_i, then each process g_{N_i} will be a_i times a birth and death process, as given above. The above differential equation is easily integrated and results also may be obtained when the birth and death processes are replaced by the corresponding diffusion approximations. When there is just one ionic species with equilibrium potential V_E and with N channels of conductance a, the steady-state mean potential is given by

$$E(V_N(\infty)) = V_E \left(\frac{\beta}{\alpha + \beta}\right) \left(\frac{\bar{G}}{G_m}\right),$$

where $\bar{G} = aN$. For further details and discussion, see Tuckwell (1987d).

Finally, we summarize some additional aspects of the present subject matter. Distributions of *sojourn times* are given in Colquhoun and Hawkes (1981). *Parameter estimation* is important, not only with a view to obtaining physical constants, but also in distinguishing various proposed models. Maximum likelihood has been one method of such estimation (Horn and Lange (1983)). Recently the problem of gaps in the recorded current which are beyond the resolving power of the measuring process has been studied (Roux and Sauve (1985)). Another associated statistical problem is that a bad choice of bin widths in waiting time histograms may lead to biased parameter estimates.

CHAPTER 9

Wiener Kernel Expansions

The method of system identification by means of analyzing the system response to white noise has been employed in neurophysiology since 1973 (references to the experiment will be given later). The mathematical basis of such a procedure will be outlined in this chapter in an expository fashion, since such accounts are not readily available. We will first attempt an elucidation of the term "system identification"; further preliminaries involve Hilbert spaces of random variables and multiple Wiener integrals.

9.1. System identification.

System identification consists of ascertaining how a physical (including biological) system works at a gross level. It consists of finding a rule from which the system responses to a class of inputs can be deduced. It does not attempt to ascertain *how* the response arises and thus seems to violate fundamental scientific principles. Nevertheless, it is not without usefulness.

As a deterministic illustration, suppose one is experimenting on a nerve cylinder upon which the potential V satisfies a linear cable equation with suitable end conditions. If the Green's function is G, then for an input current I the response with V initially zero is

$$V(x, t) = \iint G(x, y; t - s) I(y, s) \, ds \, dy.$$

Use of an impulsive input $I(x, t) = \delta(x - x_0)\delta(t)$ leads to a response $G(x, x_0; t)$, and once this is known one can ascertain the response to any input without doing an experiment. Thus the system is "identified" if G is known, and this can be measured by finding the response to an impulse. In Wiener's expansion, *white noise* plays the role of the impulse and the *Wiener kernels* play the role of G. This approach has its roots in deterministic Volterra expansions, which appeared in 1887. Note that the physical model of the nerve cylinder, consisting of all the resistive and capacitative elements that make up the cable, is no longer of interest once G is known, for we now have an input–output recipe.

Most functional neuronal networks are extremely complicated, and knowledge of their anatomical connections is usually incomplete. This fact, coupled with the complexities of nonlinear behavior of individual neurons and the uncertainties and fluctuations in numerous physiological and neurochemical variables, makes the concept of system identification appealing. We obviate the problem of fine structure and regard the network as a black box.

9.2. Hilbert spaces of random variables and multiple Wiener integrals.

In this section we introduce some concepts which are subsequently used to develop Wiener's expansion.

9.2.1. Hilbert spaces of random variables.
Let $(\Omega, \mathcal{F}, P)$ be a probability space. The set of random variables $X: \Omega \to R$ with finite second moment is a Hilbert space under the scalar product

$$(X, Y) = E(XY)$$

with corresponding norm $\|X\| = \sqrt{(X, X)}$. This Hilbert space is denoted by $L^2(\Omega)$. Two vectors in $L^2(\Omega)$ are orthogonal if

$$E(XY) = 0;$$

thus, if the two vectors have zero means, their orthogonality implies their uncorrelatedness and independence if they are also normal.

An important concept in connection with a Hilbert space is a *complete orthonormal set* (CONS) of its elements. Members of such a set are mutually orthogonal, have unit length, and any vector may be expanded in a series of them. Such a series is called a *Fourier series*. (See, e.g., Simmons (1963) for a basic treatment of Hilbert spaces.)

9.2.2. Multiple Wiener integrals.
Before discussing multiple Wiener integrals, we consider *single integrals* of the form

$$I(f) = \int_a^b f(t) \, dW(t),$$

where $[a, b] \subset \mathbb{R}$ and W is a standard Wiener process. This integral may be defined in various ways, depending on the nature and properties of f (e.g., random or not, square-integrable, etc.). If f is square-integrable with probability one, then $I(f)$ is defined as a limit of integrals of a sequence of simple functions that converges to f. When f is nonrandom, the integral has the following easily proved properties:

(i) $I(f)$ is a normal random variable with mean zero and variance given by

$$\text{Var}(I(f)) = \int_a^b f^2(t) \, dt;$$

(ii) For $f, g \in L^2[a, b]$,

$$E(I(f)I(g)) = \int_a^b f(t)g(t) \, dt.$$

From these properties we may quickly establish the following lemma.

LEMMA 9.1. *Let* $\{\phi_n(t)\}$, $n = 1, 2, \cdots$ *be a CONS for* $L^2[a, b]$. *Then* $\{I(\phi_n)\}$, $n = 1, 2, \cdots$ *is*
  (a) *a sequence of independent standard normal random variables*;
  (b) *an orthonormal set in* $L^2(\Omega)$.

Proof. For each n, $I(\phi_n)$ is a Gaussian random variable with mean zero and variance
$$\int_a^b \phi_n(t)\, dt = 1$$
by property (i) above. By property (ii) we have
$$E(I(\phi_m)I(\phi_n)) = \int_a^b \phi_m(t)\phi_n(t)\, dt = \delta_{mn}$$
by the orthonormality of the ϕ_n's. These facts will be used shortly. We now consider the multiple integrals.

For suitable real-valued nonrandom functions $f: R^p \to R$, Ito (1951a) defined a p-fold Wiener integral. In the present context, this is denoted by

(9.1) $\quad I_p(f) = \int_a^b \int_a^b \cdots \int_a^b f(t_1, t_2, \cdots, t_p)\, dW(t_1)\, dW(t_2) \cdots dW(t_p).$

Properties of this integral as needed here include the following:
  (i) If $f \in L^2[a, b]^p$, then $I_p(f)$ has mean zero and variance
$$\mathrm{Var}(I_p(f)) = p! \int_a^b \int_a^b \cdots \int_a^b |\tilde{f}(t_1, t_2, \cdots, t_p)|^2\, dt_1\, dt_2 \cdots dt_p < \infty,$$
where $\tilde{f}$ is defined as
$$\tilde{f}(t_1, t_2, \cdots, t_p) = \frac{1}{p!} \sum_{\pi(1 \cdots p)} f(t_{\pi_1}, t_{\pi_2}, \cdots, t_{\pi_p}),$$
where the sum is over all permutations of $(1, 2, \cdots, p)$. (For example, $\tilde{f}(t_1, t_2) = [f(t_1, t_2) + f(t_2, t_1)]/2!$.)
  (ii) If $p \neq q$, then $I_p(f)$ and $I_q(g)$ are orthogonal in the sense that
$$E(I_p(f)I_q(g)) = 0.$$
  (iii) $E(I_p(f)I_p(g))$
$$= p! \int_a^b \int_a^b \cdots \int_a^b \tilde{f}(t_1, t_2, \cdots, t_p)\tilde{g}(t_1, t_2, \cdots, t_p)\, dt_1\, dt_2 \cdots, dt_p.$$

Note that Ito's definition was for general Gaussian measures and not just for those induced by Brownian motion.

9.3. Cameron and Martin's expansion of an L^2-functional of a Wiener process.

In this section we will give a celebrated expansion, due to Cameron and Martin (1947), of an L^2-functional of a Wiener process in terms of what they called a *Fourier–Hermite set*. From now on we will assume $[a, b] = [0, 1]$, generalizations presenting no difficulties. We now define a set of related random variables.

DEFINITION. Let $\{\phi_n\}$ be a CONS for $L^2[0, 1]$ and put $X_n = I(\phi_n)$. Define

$$\Phi_{mp} = H_m\left(\frac{X_p}{\sqrt{2}}\right), \quad m = 0, 1, 2, \cdots, \quad p = 1, 2, \cdots,$$

where H_m is the mth of the Hermite polynomials which satisfy the orthonormality relation

$$\frac{1}{\sqrt{\pi}} \int_{-\infty}^{\infty} H_m(x) H_n(x) e^{-x^2} dx = \delta_{mn}, \quad m, n = 0, 1 \cdots.$$

Thus, Φ_{mp} is a Hermite polynomial function of the normal random variable, $X_p/\sqrt{2}$. The $\{\Phi_{mp}\}$ satisfy the following orthonormality relation.

LEMMA 9.2. *The random variables $\{\Phi_{mp}\}$ are bi-orthonormal in the sense that*

(9.2) $$(\Phi_{mp}, \Phi_{nq}) = E(\Phi_{mp}\Phi_{nq}) = \delta_{mn}\delta_{pq},$$

except when $m = n = 0$, when $(\Phi_{0p}, \Phi_{0q}) = 1$.

Proof. Consider first the case $p \neq q$. Then, as seen in Lemma 9.1, the random variables X_p and X_q are independent and, hence, so too are $H_m(X_p/\sqrt{2})$ and $H_m(X_q/\sqrt{2})$. Utilizing the fact that if X is a standard normal random variable, then $E(g(X)) = (1/\sqrt{2\pi}) \int e^{-x^2/2} g(x) dx$, we therefore have

$$E(\Phi_{mp}\Phi_{nq}) = E(\Phi_{mp})E(\Phi_{nq})$$

$$= \left[\frac{1}{\sqrt{2\pi}} \int H_m\left(\frac{x}{\sqrt{2}}\right) e^{-x^2/2} dx\right]\left[\frac{1}{\sqrt{2\pi}} \int H_n\left(\frac{y}{\sqrt{2}}\right) e^{-y^2/2} dy\right].$$

However,

$$\frac{1}{\sqrt{2\pi}} \int H_m\left(\frac{x}{\sqrt{2}}\right) e^{-x^2/2} dx = \frac{1}{\sqrt{2\pi}} \int H_0\left(\frac{x}{\sqrt{2}}\right) H_m\left(\frac{x}{\sqrt{2}}\right) e^{-x^2/2} dx,$$

which is zero, provided $m \neq 0$ by the orthogonality of the Hermite polynomials. Thus if $p \neq q$, the stated quantity vanishes except when $m = n = 0$, in which case it is unity because $H_0 = 1$.

Now consider the case $p = q$ so that both Φ_{mp} and Φ_{nq} are functions of the same random variable. Then

$$E(\Phi_{mp}\Phi_{nq}) = E\left[H_m\left(\frac{X_p}{\sqrt{2}}\right) H_n\left(\frac{X_p}{\sqrt{2}}\right)\right]$$

$$= \frac{1}{\sqrt{2\pi}} \int H_m\left(\frac{x}{\sqrt{2}}\right) H_n\left(\frac{x}{\sqrt{2}}\right) e^{-x^2/2} dx$$

$$= \delta_{mn}$$

on using the orthonormality of the Hermite polynomials. Thus the lemma is proved.

The Fourier–Hermite set consists of products of Φ_{mp}'s.

DEFINITION. The Fourier–Hermite set is the collection of random variables

$$\Psi_{m_1 m_2 \cdots m_p} = \Phi_{m_1,1} \Phi_{m_2,2} \cdots \Phi_{m_p,p},$$

where $\{m_1, \cdots, m_p\}$ is any (nonempty) set of nonnegative integers.

For example,

$$\Psi_{301} = H_3\left(\frac{X_1}{\sqrt{2}}\right) H_0\left(\frac{X_2}{\sqrt{2}}\right) H_1\left(\frac{X_3}{\sqrt{2}}\right).$$

It turns out that the Ψ's span a very large space. As their name suggests, they are an orthonormal family.

LEMMA 9.3. *The Fourier–Hermite set is orthonormal in the sense that*

$$(\Psi_{m_1 m_2 \cdots m_p}, \Psi_{n_1 n_2 \cdots n_p}) = \delta_{m_1 n_1} \delta_{m_2 n_2} \cdots \delta_{m_p n_p}.$$

Proof. The required scalar product is

$$E(\Phi_{m_1,1} \Phi_{m_2,2} \cdots \Phi_{m_p,p} \Phi_{n_1,1} \Phi_{n_2,2} \cdots \Phi_{n_p,p})$$
$$= E(\Phi_{m_1,1} \Phi_{n_1,1}) E(\Phi_{m_2,2} \Phi_{n_2,2}) \cdots E(\Phi_{m_p,p} \Phi_{n_p,p})$$

because Φ_{mp}'s with different p's are independent. The result follows from repeated use of (9.2) with $p = q$.

The main result is the following expansion, which is established in Cameron and Martin (1947).

THEOREM 9.4. *Let F be an L^2-functional of Wiener paths on $[0, 1]$ so that*

$$E(|F|^2) < \infty.$$

Put

$$F_N = \sum_{m_1, m_2, \cdots, m_N = 0}^{N} A_{m_1 m_2 \cdots m_N} \Psi_{m_1 m_2 \cdots m_N},$$

where

(9.3) $$A_{m_1 m_2 \cdots m_N} = E(F \Psi_{m_1 m_2 \cdots m_N}).$$

Then

$$\lim_{N \to \infty} E(|F_N - F|^2) = 0.$$

Thus the series of Fourier–Hermite random variables with Fourier–Hermite coefficients $\{A_{m_1 m_2 \cdots m_N}\}$ converges in mean square to an L^2-functional, F. That the coefficients are given by (9.3) follows from Lemma 9.3. We will now see how we may combine the Cameron–Martin expansion with a theorem of Ito to obtain an expansion in Wiener kernels.

9.4. Wiener kernel expansions.

In the Cameron and Martin expansion, the terms are products of Hermite polynomials of normal random variables which may be expressed as integrals of an orthonormal set in $L^2[0, 1]$. The expansion suffers from a practical point of view because as N increases, *all* the coefficients change. This difficulty can be overcome by first rewriting the terms as multiple Wiener integrals, using the following result, which is proved in Ito (1951a).

THEOREM 9.5. *Products of Hermite polynomials of the random variables* $2^{-1/2}X_v = 2^{-1/2}\int_0^1 \phi_v(t)\,dW(t)$ *can be expressed as multiple Wiener integrals as follows*:

$$(9.4) \quad \prod_{v=1}^{N} \frac{H_{m_v}(X_v/\sqrt{2})}{2^{m_v/2}} = \int_0^1 \cdots (m_1 + \cdots + m_N) \text{ times} \cdots \int_0^1 \phi_1(t_1) \cdots \phi_1(t_{m_1})$$
$$\times \phi_2(t_{m_1+1}) \cdots \phi_2(t_{m_1+m_2}) \cdots \phi_N(t_{m_1+\cdots+m_N})\,dW(t_1) \cdots dW(t_{m_1+m_2\cdots+m_N}).$$

Since $\Phi_{mp} = H_m(X_p/\sqrt{2})$, the left-hand side of (9.4) can be reexpressed as

$$\prod_{v=1}^{N} \frac{\Phi_{m_v,v}}{2^{m_v/2}} = k_{m_1\cdots m_N}\Psi_{m_1\cdots m_N},$$

where the constants k are given by

$$k_{m_1\cdots m_N} = \prod_{v=1}^{N} 2^{-m_v/2}.$$

We see, therefore, that Theorem 9.5 can be restated as

$$F = \text{l.i.m.}_{N\to\infty} \sum_{m_1,m_2,\cdots,m_N=0}^{N} \frac{A_{m_1m_2\cdots m_N}}{k_{m_1m_2\cdots m_N}} \int_0^1 \cdots (m_1 + \cdots + m_N) \text{ times} \cdots \int_0^1 \phi_1(t_1) \cdots$$
$$\phi_1(t_{m_1})\phi_2(t_{m_1+1}) \cdots \phi_N(t_{m_1+\cdots+m_N})\,dW(t_1) \cdots dW(t_{m_1+\cdots+m_N}).$$

9.4.1. Collecting terms. Finally, to obtain a Wiener-type expansion for the L^2-functional F of Brownian motion, we collect terms in the infinite series which involve the integrals of the same multiplicity.

(i) *Terms with* $\Sigma m_j = 0$. This involves only terms in which all the m_j's are zero. This term is

$$\frac{A_{00\cdots}}{k_{00\cdots}} = E(F) \doteq \mu_F.$$

(ii) *Terms with* $\Sigma m_j = 1$. Clearly, these arise when all but one of the m_j's are zero and the remaining one is unity. This involves

$$\frac{A_{100\cdots}}{k_{100\cdots}} \int \phi_1(t_1)\,dW(t_1) + \frac{A_{010\cdots}}{k_{010\cdots}} \int \phi_2(t_1)\,dW(t_1) + \cdots \doteq \int h_1(t_1)\,dW(t_1),$$

which defines h_1 as

$$h_1(t_1) = \sum_{j=1}^{\infty} a_j\phi_j(t_1), \qquad a_j = 2E\left(F\int \phi_j(t_1)\,dW(t_1)\right).$$

(iii) *Terms with* $\Sigma m_j = 2$. These terms are similar to

$$\frac{A_{0..020..}}{k_{0..020..}} \iint \phi_i(t_1)\phi_i(t_2)\,dW(t_1)\,dW(t_2),$$

where the 2 occurs in the ith subscript, and terms like
$$\frac{A_{0..010..01..}}{k_{0..010..01..}} \iint \phi_i(t_1)\phi_j(t_2)\, dW(t_1)\, dW(t_2),$$
where the 1's appear in the ith and jth places. All such terms are collected into
$$\iint h_2(t_1, t_2)\, dW(t_1)\, dW(t_2),$$
where
$$h_2(t_1, t_2) = \sum_{\substack{i=1 \\ i \leq j}}^{\infty} \sum_{j=1}^{\infty} a_{ij} \phi_i(t_1)\phi_j(t_2),$$
where
$$a_{ij} = 4\mathrm{E}\left(F \iint \phi_i(t_1)\phi_j(t_2)\, dW(t_1)\, dW(t_2)\right).$$

Note that the summation is restricted as shown because $m_i = 1$ and $m_j = 1$ can occur only once in the original expansion.

(iv) *Terms with* $\Sigma m_j = 3$. Collecting such terms together gives
$$\iiint h_3(t_1, t_2, t_3)\, dW(t_1)\, dW(t_2)\, dW(t_3),$$
where
$$h_3(t_1, t_2, t_3) = \sum_{\substack{i=1 \\ i \leq j \leq k}}^{\infty} \sum_{j=1}^{\infty} \sum_{k=1}^{\infty} a_{ijk} \phi_i(t_1)\phi_j(t_2)\phi_k(t_3),$$
$$a_{ijk} = 8\mathrm{E}\left(F \iiint \phi_i(t_1)\phi_j(t_2)\phi_k(t_3)\, dW(t_1)\, dW(t_2)\, dW(t_3)\right).$$

Similarly, we may collect terms involving quadruple integrals, etc.

In summary, we have found that
$$F = \mu_F + \int_0^1 h_1(t_1)\, dW(t_1) + \int_0^1 \int_0^1 h_2(t_1, t_2)\, dW(t_1)\, dW(t_2)$$
$$+ \int_0^1 \int_0^1 \int_0^1 h_3(t_1, t_2, t_3)\, dW(t_1)\, dW(t_2)\, dW(t_3) + \cdots,$$

where the *kernel functions* $h_1, h_2, \cdots$ are as defined above and the equality is in the sense of a mean square limit as the number of terms becomes infinite. If we put
$$g_0 = \mu_F,$$
$$g_1 = \int_0^1 h_1(t_1)\, dW(t_1),$$
$$g_2 = \int_0^1 \int_0^1 h_2(t_1, t_2)\, dW(t_1)\, dW(t_2),$$
$$\vdots$$
$$g_k = \int_0^1 \cdots k \text{ times} \cdots \int_0^1 h_k(t_1, \cdots, t_k)\, dW(t_1) \cdots dW(t_k),$$

then we see that

(9.5) $$F = \sum_{k=0}^{\infty} g_k,$$

and since g_k's for different k involve multiple Wiener integrals of different orders, they are mutually orthogonal:

$$(g_j, g_k) = 0, \quad j \neq k.$$

The expansion (9.5) is similar to what is generally known as a Wiener kernel expansion. In order to obtain an expansion that corresponds exactly with Wiener's (1958), we first note that multiple Wiener integrals can be expressed as iterated stochastic integrals. For example,

(9.6) $$\int_0^1 \int_0^1 f(t_1, t_2) \, dW(t_1) \, dW(t_2) = 2! \int_0^1 \left(\int_0^{t_2} \tilde{f}(t_1, t_2) \, dW(t_1) \right) dW(t_2).$$

Since the integrand for the $dW(t_2)$ integration is a random process, the integral depends on the definition of a stochastic integral. It is easy to show that Ito's definition of multiple Wiener integral corresponds to the use of the Ito form of the stochastic integral in (9.6) and, hence, the expansion given above.

To see what results if we use a Stratonovich integral (see, e.g., Davis (1978)), we use the general definition

$$\int_0^t Y(s) \, dX(s) = \int_0^t Y(s) \, dX(s) + \langle X, Y \rangle(t),$$

    Stratonovich   Ito,

where X, Y are continuous local semimartingales and $\langle X, Y \rangle(t)$ is the quadratic covariance of X and Y over $[0, t]$. A calculation from first principles with

$$Y(s) = \int_0^s f(u, s) \, dW(u), \quad X(s) = W(s)$$

gives the required covariance process and results in the relation

$$2! \int_0^1 \left(\int_0^{t_2} \tilde{f}(t_1, t_2) \, dW(t_1) \right) dW(t_2) = 2! \int_0^1 \int_0^{t_2} \tilde{f}(t_1, t_2) \, dW(t_2) - \int_0^1 \tilde{f}(t_1, t_1) \, dt_1 \quad \text{a.s.}$$

   Ito   Stratonovich

The first three terms in (9.5) can then be written

$$g_0 = E(F),$$

$$g_1 = \int_0^1 h_1(t_1) \, dW(t_1),$$

$$g_2 = \int_0^1 \int_0^1 h_2(t_1, t_2) \, dW(t_1) \, dW(t_2) - \int_0^1 h_2(t_1, t_1) \, dt,$$

as in Wiener's expansion. The next term is

$$g_3 = \int_0^1 \int_0^1 \int_0^1 h_3(t_1, t_2, t_3) \, dW(t_1) \, dW(t_2) \, dW(t_3)$$

$$- 3 \int_0^1 \int_0^1 h_3(t_1, t_1, t_2) \, dt_1 \, dW(t_2).$$

9.5. Measurement of the kernel functions and experimental results.

We will briefly describe a method which is commonly employed to obtain the Wiener kernels $h_1, h_2, \cdots$. Initially the method was suggested by Lee and Schetzen (1965) and is elaborated on in Marmarelis and Marmarelis (1978) and Schetzen (1980). We adopt the view that a system, possibly nonlinear, may be characterized by its Wiener kernels. "Identification" will consist of determining the Wiener kernels.

Let a system response to a Gaussian white noise $w(t)$ be

$$y(t) = \text{l.i.m.}_{N \to \infty} \sum_{k=0}^{N} G_k,$$

where the G_k's are random functionals

$$G_0 = E(y(t)),$$

$$G_1 = \int_{-\infty}^{\infty} h_1(\tau_1) w(t - \tau_1) \, d\tau_1,$$

$$G_2 = \int_{-\infty}^{\infty} \int_{-\infty}^{\infty} h_2(\tau_1, \tau_2) w(t - \tau_1) w(t - \tau_2) \, d\tau_1 \, d\tau_2$$

$$- K \int_{-\infty}^{\infty} h_2(\tau_2, \tau_2) \, d\tau_2$$

$$\vdots$$

where $\text{Cov}(w(t_1), w(t_2)) = K\delta(t_1 - t_2)$. One now assumes *ergodicity* so that G_0 can be estimated from

$$\hat{G}_0 = \frac{1}{2T} \int_{-T}^{T} y(t) \, dt.$$

To obtain the next kernel, h_1, one obtains a "delayed version" $w(t - \sigma)$ of white noise that is considered to be a functional of first order because it can be written

$$w(t - \sigma) = \int_{-\infty}^{\infty} \delta(t - \sigma) w(t - \tau) \, d\tau.$$

This means it is orthogonal to any functional of order other than one, and this yields

$$E(y(t)w(t-\sigma)) = E(G_1 w(t-\sigma))$$
$$= \left(\int_{-\infty}^{\infty} h_1(\tau_1) w(t-\tau_1) w(t-\sigma) \, d\tau_1\right)$$
$$= K \int_{-\infty}^{\infty} h_1(\tau_1) \delta(\tau_1 - \sigma) \, d\tau_1$$
$$= K h_1(\sigma).$$

Thus

$$\hat{h}_1(\sigma) = \frac{K^{-1}}{2T} \int_{-T}^{T} y(t) w(t-\sigma) \, dt.$$

Most experimental studies attempting system identification in neurobiology have been on visual systems. (Marmarelis and Naka (1973), McCann (1974), Naka, Sakuranaga, and Ando (1985), Sakuranaga and Naka (1985)). It is claimed that the kernels up to h_2 give a good approximation in computing the system response.

The theory of the kinds of expansions presented in this chapter has been extended in several directions. Compact derivations of some of the above results are in Hida (1980) and Kallianpur (1980). Expansions for functionals of Poisson processes were developed by Ogura (1972) and for independent-increment processes by Segall and Kailath (1976). Other theoretical investigations are those of McKean (1973), Yasui (1979), and Isobe and Sato (1984). Park (1970) contains a multiparameter version of Cameron and Martin's orthogonal development, and expansions involving multiparameter Wiener processes are also given in Hajek and Wong (1981). There seems to be a wide gap between the mathematical developments and the applications, which is perhaps not surprising considering the mathematical complexities.

CHAPTER 10

The Stochastic Activity of Neuronal Populations

We have examined processes connected with the stochastic activity of neurons and of ionic channels whose dimensions are much smaller. In the opposite direction in vertebrate nervous systems are aggregates of neurons and glia which may contain millions and perhaps billions of cells. Some examples are the various components of the cerebral hemispheres and substructures such as thalamic nuclei, the hippocampus, and the cerebellum. Understanding the activities of such collections of cells, which are often called *neural networks,* is clearly of great interest, especially in relation to behavior, cognition, and perception. There is also clinical interest in the effects of abnormalities and pathologies of the nervous system, and the first part of our present brief discussion will be relevant to this aspect. Thus we will be concerned with neurophysiological phenomena rather than with psychological ones. We will first describe methods for the analysis of evoked potentials, then of the EEG. We will also quickly review some approaches to the modeling of populations of cells.

10.1. Evoked potentials.

We may define an *evoked potential* (EP) as an electrical response of a nervous system to a sensory stimulus. However, sometimes other stimuli are employed, such as electrical shocks to certain nerve fibers, stimuli to muscles, etc. The responses may be observed in a single cell or grossly from several cells. A *visually evoked potential* might be a response to a light flash, for example. An *auditory evoked potential* may result from stimulation by tones.

In human studies such EPs are usually recorded from the scalp in the same manner as the EEG. Those generated in the primary sensory areas (e.g., the occipital cortex for visual EPs) have latencies of 10–20 msec and amplitudes of up to $10\,\mu V$. It is probable that EPs are a manifestation of EPSPs and IPSPs that arise as a consequence of the sensory stimulus and not action potentials Bazar (1980)). One clinical use of EPs is in the nonsurgical detection of lesions in sensory pathways.

10.1.1. Analysis and estimation of EPs. In a recording of an EP much noise is present due to the background EEG, electrical interference, muscle artifacts, etc. The amplitude of the noise may be much greater than that of the EP so that EPs are always averaged over many trials. In fact, EP means an *averaged* EP. However, a large number of trials must be performed to reduce the confidence limits of the noise to acceptable values, and this is very inconvenient and inefficient.

Wiener filtering. The theory of Wiener filtering (Wiener (1949)) has been elaborated on for discrete and continuous weakly stationary processes by several authors (e.g., Yaglom (1962), Papoulis (1977)). A rigorous treatment can be found in Wong (1971 and later editions).

DEFINITION. Let $\{X(t)\}$, $t \in T$, be a real second-order process, and let Y be a real second-order random variable. Let $\mathcal{H}_X$ be the Hilbert space generated by $\{X(t)\}$. A random variable $\tilde{Y} \in \mathcal{H}_X$ satisfying

$$E|\tilde{Y} - Y|^2 = \inf_{Z \in \mathcal{H}_X} E|Z - Y|^2$$

is called a linear least-squares estimator of Y given $\{X(t)\}$.

Wong (1971) proved that the following gives a necessary and sufficient condition for a random variable to be such an estimator.

PROPOSITION 10.1. $\hat{Y}$ *is a linear least-squares estimator of* Y *given* $\{X(t)\}$ *if and only if*

$$E[(\hat{Y} - Y)X(t)] = 0$$

for all $t \in T$.

In the estimation problem for EPs, it has always been assumed that the *noise* $N(t)$ is additive. If the *signal* is $S(t)$, then what is *observed* is the sum

$$X(t) = S(t) + N(t).$$

It is required to estimate $S(t)$ from a recording of $X(t)$. The key assumption is that a relation of the following kind exists between the estimator $\hat{S}(t)$ and $X(t)$:

$$\hat{S}(t) = \int X(t - \alpha)h(\alpha)\,d\alpha.$$

Thus if one can find the function (filter) h, then one can determine $\hat{S}$ by integrating the observed record times h.

The standard theory entails several assumptions, including the following:
(a) the random processes $S(t)$ and $N(t)$ are weakly stationary;
(b) records are available from $t = -\infty$ to $t = +\infty$;
(c) the noise has zero mean;
(d) $S(t)$ and $N(t)$ are uncorrelated so that

$$E(S(t_1)N(t_2)) = E(S(t_1))E(N(t_2)) \quad \forall t_1, t_2.$$

Then it is found that $\hat{S}(t)$ is a linear least-squares estimator for $S(t)$ if the

Fourier transform of h,

$$H(\omega) = \int_{-\infty}^{\infty} h(t)e^{-i\omega t}\, dt,$$

is given by

(10.1) $$H(\omega) = \frac{\Phi_{SS}(\omega)}{\Phi_{SS}(\omega) + \Phi_{NN}(\omega)},$$

where Φ_{NN} and Φ_{SS} are the spectral densities of the noise and signal processes, respectively.

Whereas the above formulation solves the problem of estimating $S(t)$, it does so only when the spectral densities of signal and noise are known. There have been some intriguing attempts to apply formula (10.1) when these spectral densities are not given. Walter (1969) suggested that if there were n trials, then an estimate of $H(\omega)$ could be obtained from

$$\hat{H}(\omega) = \frac{(n/(n-1))\Phi_{\bar{X}\bar{X}}(\omega) - (1/(n-1))\overline{\Phi_{XX}(\omega)}}{\Phi_{XX}(\omega)},$$

where $\Phi_{\bar{X}\bar{X}}$ is the spectrum of the average EP and $\overline{\Phi_{XX}}$ is the average of the spectra of the individual EPs. Doyle (1975) claimed that the correct version is

$$\hat{H}(\omega) = \frac{(n/(n-1))\Phi_{\bar{X}\bar{X}}(\omega) - (1/(n-1))\overline{\Phi_{XX}(\omega)}}{\Phi_{\bar{X}\bar{X}}(\omega)}$$

and this discrepancy needs further attention. Carlton and Katz (1980) discuss the merits of Wiener filtering of EPs. An alternative procedure has been proposed by Brillinger (1978), (1981). The response to a single stimulus is $s(t)$, which is unknown but fixed. The background noise is a stationary process $\{\epsilon(t)\}$, and stimuli arrive at event times in a counting process $\{M(t)\}$. The whole evoked-response record may then be written

$$Y(t) = \int s(t-u)\, dM(u) + \epsilon(t), \quad t \in [0, T].$$

The standard estimator for $s(t)$ is

$$\hat{S}(t) = \frac{1}{M(T)} \int_0^{T-t} Y(t+u)\, dM(u).$$

This estimator is biased, but under certain conditions it is asymptotically normal. As an alternative, Brillinger provides estimators of the Fourier transform of $s(t)$, but such a method does not yet seem to have been applied.

10.2. EEG.

An EEG is a recording of electrical potential as a function of time, made either from the scalp or from the brain surface (electrocorticogram, or ECoG). In scalp recordings, one electrode is usually placed in a "neutral" area such as an ear lobe. The amplitudes are in the range 10–200 μV.

Various frequency bands (Hz) are given Greek letter assignments: δ, 0.5–3.5; θ, 3–8; α, 8–13; and β, 14–30. The pattern of the EEG depends strongly on conditions.

The fluctuations in the scalp EEG reflect, after much attenuation, the activity in nearby neuronal populations. It has been deduced that the most likely source of the fluctuations is the occurrence of occasional bursts of postsynaptic potentials in (for example) pyramidal cells, perhaps as a consequence of activity in thalamocortical circuits (Creutzfeldt (1974)).

One distinguishes epochs of EEG records that are approximately stationary from those that are manifestly nonstationary. According to several investigators, stationary EEGs are well approximated by zero mean Gaussian processes (see Johnson, Wright, and Segall (1979)). For stationary recordings the standard statistical descriptions may be given, including *autocovariance*, associated *spectral density*, and *cross-covariance*. The computing of spectral densities has been facilitated by the *fast Fourier algorithm*. To further reduce computational labor, *Walsh transforms* (and fast Walsh transforms) have been employed, but this has been met with favor by some (Larsen and Lai (1980), Dzwonczyk, Howie, and McDonald (1984)) and not by others (Barlow (1979), Jansen, Bourne, and Ward (1981)).

Perhaps the most significant problems lie in detecting and classifying nonstationary epochs. Examples include changes due to various sleep stages, anesthesia, and pathological conditions such as brain tumors or epilepsy. There are trained electroencephalographers who can distinguish various modes, but there is a need for automatic classification. Useful attacks have been with *compressed spectral arrays* (power spectra versus time), *autoregressive time series* models (e.g., Gersch, Yonemoto, and Naitch (1977)), and *Kalman filtering* (Bolin (1977)).

10.3. Stochastic neuroanatomy.

Accurate quantitative data on neuronal populations are necessary if any model for the behavior of such a population can be tested. Unfortunately, such data is rare due to the painstaking nature of the work involved. See Blinkov and Glezer (1968) for some relevant data.

Important data include the following:

(a) the number density of different types of neurons and glia and their spatial variations;

(b) the size distributions of dendritic trees and cell bodies as well as axon diameters;

(c) extracellular space;

(d) number densities of synapses of various kinds together with geometric details of the connections made by cells of a given type with cells of other types;

(e) morphology of individual cell types.

Theoretical models that enable us to estimate these quantities from limited data are important. An early attempt was that of Uttley (1956), who studied anatomical interactions amongst dendritic and axonal populations. Takizawa and Oonuki (1980) addressed the estimation of synapse densities and emphasized the fact that neuronal data is biased towards larger cells. Note that if a convenient origin is chosen (such quantities as $N(x, y, z)$), the number of synapses in $[0, x] \times [0, y] \times [0, z]$, is a three-parameter stochastic process.

Networks of randomly connected model neurons have attracted some attention. For example, Grenander and Silverstein (1977) considered cell populations that may be divided into two classes of neurons with randomly assigned connections. A study of the spectral properties of associated linear operators gave rise to a law of large numbers, wherein the influence of the randomness in topology approached zero as the network grew in size. In the same context, it is possible that results such as the following from random graph theory (Erdös and Rényi (1960)) may be relevant.

THEOREM 10.2. *Consider a random graph with n vertices. If the number of edges is*
$$N \sim \rho n^{k-2/k-1},$$
then the number N_k of isolated trees of order k has the asymptotic law
$$\Pr\{N_k = j\} \xrightarrow{d} \frac{\lambda^j e^{-\lambda}}{j!}, \qquad j = 0, 1, \cdots,$$
where
$$\lambda = \frac{(2\rho)^{k-1} k^{k-2}}{k!}.$$

In the present context a vertex represents a connection between two nerve cells. For a comprehensive treatment of random graphs, see Bollobás (1985).

10.4. Dynamical models.

There have been several attempts to construct models of the ongoing activity of populations of neurons. These include randomly connected networks of McCulloch–Pitts-type model neurons (see Griffiths (1971)) and the models of Beurle (1956) and Wilson and Cowan (1972), (1973). In general, stochastic effects are not addressed in obtaining results, but Venzl (1976) added Gaussian white noise to the kinetic equations (space-clamped) of Wilson and Cowan to obtain a two-dimensional Markov process:

$$d\begin{bmatrix} E \\ I \end{bmatrix} = \begin{bmatrix} -E + \tfrac{1}{2}S_1 & (E - \gamma_1 I + p_1) \\ -I + \tfrac{1}{2}S_2 & (\gamma_2 E - I + p_2) \end{bmatrix} dt + c \begin{bmatrix} dW_1 \\ dW_2 \end{bmatrix}.$$

Here E and I are averaged proportions of active excitatory and inhibitory neurons; S_1 and S_2 are sigmoidal functions; p_1, p_2, γ_1, γ_2 are neurophysiological constants; and W_1, W_2 are independent standard Wiener processes. Venzl obtained, under simplifying assumptions, various statistical properties of the steady-state distribution and correlation functions.

There have also been attempts to borrow ideas from statistical mechanics to construct neural population models (Cowan (1970), Ingber (1982)). An excellent review of network modeling was given by Levine (1983). It is difficult, in constructing such models, to know what neurobiological variables should be included and which dropped without sacrificing too many actual properties of real neuronal populations.

References

J. ABRAHAMS (1985), *A survey of recent progress on level-crossing problems for random processes*, in Communications and Networks: A Survey of Recent Advances, I. F. Blake and H. V. Poor, eds., Springer-Verlag, Berlin, New York.

M. ABRAMOWITZ AND I. STEGUN (EDS.) (1965), *Handbook of Mathematical Functions*, Dover, New York.

J. S. BARLOW (1979), *Computerized clinical electroencephalography in perspective*, IEEE Trans. Biomed. Engrg., BME-26, pp. 377–391.

J. N. BARRETT AND W. E. CRILL (1974), *Specific membrane properties of cat motoneurones*, J. Physiol., 239, pp. 301–324.

M. S. BARTLETT (1963), *The spectral analysis of point processes*, J. Roy. Statist. Soc. B, 25, pp. 264–295.

E. BAZAR (1980), *EEG-Brain Dynamics*, Elsevier, Amsterdam.

G. A. BECUS (1977), *Random generalized solutions to the heat equation*, J. Math. Anal. Appl., 60, pp. 93–102.

―――― (1978), *Solution to the random heat equation by the method of successive approximations*, J. Math. Anal. Appl., 64, pp. 277–296.

M. R. BENNETT AND T. FLORIN (1974), *A statistical analysis of the release of acetylcholine at newly formed synapses in striated muscle*, J. Physiol., 238, pp. 93–107.

H. BERGER (1929), *Uber das elektrenkaphalogramm des menschen*, Arch. Psych. Nerven, 87, pp. 527–570.

R. L. BEURLE (1956), *Properties of a mass of cells capable of regenerating pulses*, Philos. Trans. Roy. Soc. A, 240, pp. 55–97.

P. BILLINGSLEY (1968), *Convergence of Probability Measures*, John Wiley, New York.

P. O. BISHOP, W. R. LEVICK, AND W. O. WILLIAMS (1964), *Statistical analysis of the dark discharge of lateral geniculate neurons*, J. Physiol., 170, pp. 598–612.

E. A. BLAIR AND J. ERLANGER (1932), *Responses of axons to brief shocks*, Proc. Soc. Exper. Biol. Med., 29, pp. 926–927.

I. F. BLAKE AND W. C. LINDSEY (1973), *Level-crossing problems for random processes*, IEEE Trans. Inform. Theory, IT-19, pp. 295–315.

S. M. BLINKOV AND I. T. GLEZER (1968), *The Human Brain in Figures and Tables*, Plenum Press, New York.

G. W. BLUMAN AND H. C. TUCKWELL (1987), *Techniques for obtaining analytical solutions for Rall's model neuron*, J. Neurosci. Methods, 20, pp. 151–166.

T. BOLIN (1977), *Analysis of EEG signals with changing spectra using a short-word Kalman estimator*, Math. Biosci., 35, pp. 221–259.

B. BOLLOBÁS (1985), *Random Graphs*, Academic Press, New York.

I. A. BOYD AND A. R. MARTIN (1956), *The end-plate potential in mammalian muscle*, J. Physiol., 132, pp. 74–91.

REFERENCES

D. R. BRILLINGER (1975), *The identification of point process systems*, Ann. Probab., 3, pp. 909–929.

────── (1978), *A note on the estimation of evoked responses*, Biol. Cybernet., 31, pp. 141–144.

────── (1981), *The general linear model in the design and analysis of evoked response experiments*, J. Theoret. Neurobiol., 1, pp. 105–119.

────── (1987), *Analysing interacting nerve cell spike trains to assess causal connections*, in Advanced Methods of Physiological System Modeling, Vol. 1, V. Z. Marmarelis, ed., Biomedical Simulations Resource, University of Southern California, Los Angeles.

────── (1988), *Some statistical methods for random process data from seismology and neurophysiology*, Ann. Statist., 16, pp. 1–54.

F. BRINK, D. W. BRONK, AND M. B. LARRABEE (1946), *Chemical excitation of nerve*, Ann. New York Acad. Sci., 47, pp. 457–485.

T. H. BROWN, D. H. PERKEL, AND M. W. FELDMAN (1976), *Evoked neurotransmitter release: Statistical effects of nonuniformity and nonstationarity*, Proc. Nat. Acad. Sci. U.S.A., 73, pp. 2913–2917.

H. L. BRYANT, A. R. MARCOS, AND J. P. SEGUNDO (1973), *Correlations of neuronal spike discharges produced by monosynaptic connections and by common inputs*, J. Neurophysiol., 36, pp. 205–225.

H. L. BRYANT AND J. P. SEGUNDO (1976), *Spike initiation by transmembrane current: A white noise analysis*, J. Physiol., 260, pp. 279–314.

W. BUNO, J. FUENTES, AND J. P. SEGUNDO (1978), *Crayfish stretch-receptor organs: Effects of length steps with and without perturbations*, Biol. Cybernet., 31, pp. 99–110.

B. D. BURNS (1968), *The Uncertain Nervous System*, Arnold, London.

B. D. BURNS AND A. C. WEBB (1976), *The spontaneous activity of neurones in the cat's cerebral cortex*, Proc. Roy. Soc. London Ser. B, 194, pp. 211–222.

W. H. CALVIN AND C. F. STEVENS (1965), *A Markov process model for neuron behavior in the interspike interval*, Proc. 18th Annual Conference on Engineering in Medicine and Biology, 7, 118, Abstract.

────── (1968), *Synaptic noise and other sources of randomness in motoneuron interspike intervals*, J. Neurophysiol., 31, pp. 574–587.

R. H. CAMERON AND W. T. MARTIN (1947), *The orthogonal development of nonlinear functionals in series of Fourier–Hermite functionals*, Ann. of Math., 48, pp. 385–392.

R. M. CAPOCELLI AND L. M. RICCIARDI (1971), *Diffusion approximation and first passage time problem for a model neuron*, Kybernetik, 8, pp. 214–223.

E. H. CARLTON AND S. KATZ (1980), *Is Wiener filtering an effective method of improving evoked potential estimation?*, IEEE Trans. Biomed. Engrg., BME-27, pp. 187–192.

S. CHANDRESEKHAR (1943), *Dynamical friction II. The rate of escape of stars from clusters and the evidence for the operation of dynamical friction*, Astrophys. J., 97, pp. 263–273.

C. CHATFIELD (1980), *The Analysis of Time Series: Theory and Practice*, Chapman and Hall, London.

E. CINLAR (1972), *Superposition of point processes*, in Stochastic Point Processes: Statistical Analysis, Theory and Applications, P. A. W. Lewis, ed., John Wiley, New York.

J. R. CLAY (1976), *A stochastic analysis of the graded excitatory response of nerve membrane*, J. Theoret. Biol., 59, pp. 141–158.

D. COLQUHOUN AND A. G. HAWKES (1977), *Relaxation and fluctuations of membrane currents that flow through drug-operated channels*, Proc. Roy Soc. Ser. B., 199, pp. 231–262.

────── (1981), *On the stochastic properties of single ion channels*, Proc. Roy. Soc. Ser. B, 211, pp. 205–235.

────── (1982), *On the stochastic properties of bursts of single ion channel openings and of clusters of bursts*, Philos. Trans. Roy. Soc. B, 300, pp. 1–59.

S. CONRADI (1969), *On motoneuron synaptology in adult cats*, Acta. Physiol. Scand. Suppl., 332.

D. K. COPE AND H. C. TUCKWELL (1979), *Firing rates of neurons with random excitation and inhibition*, J. Theoret. Biol., 80, pp. 1–14.

M. J. CORREIA AND J. P. LANDOLT (1979), *A point process analysis of the spontaneous activity of anterior semicircular canal units in the anesthetized pigeon*, Biol. Cybernet., 27, pp. 199–213.

J. D. COWAN (1970), *A statistical mechanics of nervous activity* in Lectures on Mathematics in the Life Sciences, Vol. 2, M. Gerstenhaber, ed., American Mathematical Society, Providence, RI.

D. R. COX (1962), *Renewal Theory*, Methuen, London.

D. R. COX AND D. V. HINKLEY (1975), *Theoretical Statistics*, Chapman and Hall, London.

D. R. COX AND P. A. W. LEWIS (1966), *The Statistical Analysis of Series of Events*, Methuen, London.

D. R. COX AND H. D. MILLER (1965), *The Theory of Stochastic Processes*, John Wiley, New York.

H. CRAMER AND M. R. LEADBETTER (1967), *Stationary and Related Stochastic Processes*, John Wiley, New York.

O. CREUTZFELDT (1974), *The neuronal generation of the EEG*, in Handbook of Electroencephalography and Clinical Neurophysiology, Vol. 2, Part C, A. Redmond, ed., Elsevier, Amsterdam.

R. F. CURTAIN (1977), *Stochastic evolution equations with general white noise disturbance*, J. Math. Anal. Appl., 60, pp. 570–595.

R. F. CURTAIN AND P. L. FALB (1971), *Stochastic differential equations in Hilbert space*, J. Differential Equations, 10, pp. 412–430.

H. H. DALE, W. FELDBERG, AND M. VOGT (1936), *Release of acetylcholine at voluntary motor nerve endings*, J. Physiol., 86, pp. 353–380.

D. A. D. DARLING AND A. J. F. SIEGERT (1953), *The first passage time problem for a continuous Markov process*, Ann. Math. Statist., 24, pp. 624–639.

—— (1957), *A systematic approach to a class of problems in the theory of noise and other random phenomena. Part I*, IRE Trans. Inform. Theory, 3, pp. 32–37.

M. H. A. DAVIS (1978), *Martingale integrals and stochastic calculus*, in Communication Systems and Random Process Theory, J. K. Skwirzynski, ed., Sijthoff and Noordhoff, Alphen aan den Rijn.

J. DEL CASTILLO AND B. KATZ (1955), *Local activity at a depolarized nerve-muscle junction*, J. Physiol., 128, pp. 396–411.

—— (1957), *Interaction at end-plate receptors between different choline derivatives*, Proc. Roy. Soc. Ser. B., 146, pp. 369–381.

G. J. DINNING AND A. C. SANDERSON (1981), *Real-time classification of multiunit neural signals using reduced feature sets*, IEEE Trans. Biomed. Engrg., BME-28, pp. 804–808.

C. R. DOERING (1985), *Nonperturbative bounds on $\langle \phi^n \rangle$ in cutoff $(\lambda \phi^n)_d$ field theory*, Phys. Rev. Lett., 55, pp. 1657–1660.

J. DOOB (1942), *The Brownian movement and stochastic equations*, Ann. of Math., 43, pp. 351–369.

D. J. DOYLE (1975), *Some comments on the use of Wiener filtering for the estimation of evoked potentials*, EEG Clinical Neurophysiol., 38, pp. 533–534.

R. DZWONCZYK, M. B. HOWIE, AND J. S. MCDONALD (1984), *A comparison between Walsh and Fourier analysis of the EEG for tracking the effects of anesthesia*, IEEE Trans. Biomed. Engrg., BME-31, pp. 551–556.

J. C. ECCLES (1957), *The Physiology of Nerve Cells*, Johns Hopkins University Press, Baltimore, MD.

—— (1964), *The Physiology of Synapses*, Springer-Verlag, Berlin, New York.

—— (1984), *The Human Mystery*, Routledge, London.

A. EKHOLM AND J. HYVARINEN (1970), *A pseudo-Markov model for series of neuronal spike events*, Biophys. J., 10, pp. 773–796.

P. S. ENGER, J. K. S. JANSEN, AND L. WALLOE (1969), *A biological model of the excitation of a second order sensory neurone*, Kybernetik, 6, pp. 141–145.

P. ERDÖS AND A. RÉNYI (1960), *On the evolution of random graphs*, Publ. Math. Inst. Hungarian Acad. Sci., 5, pp. 17–61.

S. N. ETHIER AND T. G. KURTZ (1986), *Markov Processes: Characterization and Convergence*, John Wiley, New York.

W. G. FARIS AND G. JONA-LASINIO (1982), *Large fluctuations for a nonlinear heat equation with noise*, J. Phys. A, 15, pp. 3025–3055.

P. FATT AND B. KATZ (1950), *Some observations on biological noise*, Nature, 166, pp. 597–598.

——— (1952), *Spontaneous subthreshold activity at motor nerve endings*, J. Physiol., 117, pp. 109–128.

P. D. FEIGIN (1976), *Maximum likelihood estimation for continuous-time stochastic processes*, Adv. Appl. Probab., 8, pp. 712–736.

W. FELLER (1966), *An Introduction to Probability Theory and Its Applications*, Vol. 2, John Wiley, New York.

R. D. FERNALD (1971), *A neuron model with spatially distributed synaptic input*, Biophys. J., 11, pp. 323–340.

A. S. FINKEL AND S. J. REDMAN (1983), *The synaptic current evoked in cat spinal motoneurones by impulses in single group Ia axons*, J. Physiol., 342, pp. 615–632.

R. FITZHUGH (1961), *Impulses and physiological states in theoretical models of nerve membrane*, Biophys. J., 1, pp. 445–466.

E. FLORATOS AND J. ILIOPOULOS (1983), *Equivalence of stochastic and canonical quantization in perturbation theory*, Nuclear Phys. B, 214, pp. 392–404.

K. FRANK AND M. G. F. FUORTES (1955), *Potentials recorded from the spinal cord with microelectrodes*, J. Physiol., 130, pp. 625–654.

P. FRANKEN (1963), *A refinement of the limit theorem for the superposition of independent renewal processes*, Teor. Veroyatnost. i Primenen, 8, pp. 320–328.

B. FRANKENHAEUSER AND A. F. HUXLEY (1964), *The action potentials of the myelinated nerve fibre of Xenopus laevis as computed on the basis of voltage clamp data*, J. Physiol., 171, pp. 302–315.

C. D. GEISLER AND J. M. GOLDBERG (1966), *A stochastic model of the repetitive activity of neurons*, Biophys. J., 6, pp. 53–69.

W. GERSCH, J. YONEMOTO, AND P. NAITOH (1977), *Automatic classification of multivariate EEGs*, Comput. Biomed. Res., 10, pp. 297–318.

G. L. GERSTEIN (1962), *Mathematical models for the all or none activity of some neurons*, IRE Trans. Inform. Theory, 8, pp. 137–143.

G. L. GERSTEIN AND N. Y-S. KIANG (1960), *An approach to the quantitative analysis of electro-physiological data from signal neurons*, Biophys. J., 1, pp. 15–28.

G. L. GERSTEIN AND B. MANDELBROT (1964), *Random walk models for the spike activity of a single neuron*, Biophys. J., 4, pp. 41–68.

I. I. GIHMAN AND A. V. SKOROHOD (1972), *Stochastic Differential Equations*, Springer-Verlag, Berlin, New York.

B. GLUSS (1967), *A model for neuron firing with exponential decay of potential resulting in diffusion equations for probability density*, Bull. Math. Biophys., 29, pp. 233–243.

J. M. GOLDBERG, H. O. ADRIAN, AND F. D. SMITH (1964), *Response of neurons of the superior olivary complex of the cat to acoustic stimuli of long duration*, J. Neurophysiol., 27, pp. 706–749.

I. S. GRADSHTEYN AND I. M. RYZHIK (1965), *Table of Integrals Series and Products*, Academic Press, New York.

U. GRENANDER AND J. W. SILVERSTEIN (1977), *Spectral analysis of networks with random topologies*, SIAM J. Appl. Math., 32, pp. 499–519.

J. S. GRIFFITHS (1971), *Mathematical Neurobiology*, Academic Press, New York.

B. GRIGELIONIS (1963), *On the convergence of sums of random step processes to a Poisson process*, Teor. Veroyatnost. i Primenen, 8, pp. 189–194. (English translation, in Theory. Probab. Appl., 8, pp. 172–182.)

D. V. GUSAK AND V. S. KORALYUK (1968), *On the first passage time across a given level for processes with independent increments*, Theory. Probab. Appl., 13, pp. 448–456.

R. GUTTMAN, L. FELDMAN, AND H. LECAR (1974), *Squid axon membrane response to white noise stimulation*, Biophys. J., 14, pp. 941–955.

M. HABIB AND P. K. SEN (1985), *Non-stationary stochastic point-process models in neurophysiology with applications to learning*, in Biostatistics: Statistics in Biomedical, Public Health and Environmental Sciences, P. K. Sen, ed., Elsevier, Amsterdam.

S. HAGIWARA (1954), *Analysis of interval fluctuation of the sensory nerve impulse*, Japan J. Physiol., 4, pp. 234–240.

B. HAJEK AND E. WONG (1981), *Set-parametered martingales and multiple stochastic integration*, in Stochastic Integrals, D. Williams, ed., Springer-Verlag, Berlin, New York.

F. B. HANSON AND H. C. TUCKWELL (1983), *Diffusion approximations for neuronal activity including synaptic reversal potentials*, J. Theoret. Neurobiol., 2, pp. 127–153.

T. HIDA (1980), *Brownian Motion*, Springer-Verlag, Berlin, New York.

B. HILLE (1984), *Ionic Channels of Excitable Membranes*, Sinauer, Sunderland, MA.

A. L. HODGKIN AND A. F. HUXLEY (1939), *Action potentials recorded from inside a nerve fibre*, Nature, 144, pp. 710–711.

—— (1952), *A quantitative description of membrane current and its application to conduction and excitation nerve*, J. Physiol., 117, pp. 500–544.

A. L. HODGKIN AND W. A. H. RUSHTON (1946), *The electrical constants of a crustacean nerve fibre*, Proc. Roy. Soc. London Ser. B, 111, pp. 175–188.

A. V. HOLDEN (1976), *Models of the Stochastic Activity of Neurones*, Springer-Verlag, Berlin, New York.

F. HOOGE (1976), *1/f noise*, Physica, 83B, pp. 14–23.

R. HORN AND K. LANGE (1983), *Estimating kinetic constants from single channel data*, Biophys. J., 43, pp. 207–223.

D. L. IGLEHART (1965), *Limiting diffusion approximations for the many server queue and the repairman problem*, J. Appl. Probab., 2, pp. 429–441.

L. INGBER (1982), *Statistical mechanics of neocortical interactions*, Physica, 5D, pp. 83–107.

E. ISOBE AND S. SATO (1984), *An integro-differential formula on the Wiener kernels and its application to sandwich system identification*, IEEE Trans. Automat. Control, AC-29, pp. 595–602.

K. ITO (1951a), *Multiple Wiener integral*, J. Math. Soc. Japan, 3, pp. 157–169.

—— (1951b) *On stochastic differential equations*, Mem. Amer. Math. Soc., 4.

—— (1984), *Infinite dimensional Ornstein–Uhlenbeck processes*, in Stochastic Analysis, North-Holland, Amsterdam, pp. 197–224.

J. J. B. JACK, D. NOBLE, AND R. W. TSIEN (1985), *Electric Current Flow in Excitable Cells*, Clarendon, Oxford.

J. J. B. JACK AND S. J. REDMAN (1971), *An electrical description of the motoneurone and its application to the analysis of synaptic potentials*, J. Physiol., 215, pp. 321–352.

M. B. JACKSON (1985), *Stochastic behavior of a many-channel membrane system*, Biophys. J., 47, pp. 129–137.

B. H. JANSEN, J. R. BOURNE, AND J. W. WARD (1981), *Spectral decomposition of EEG intervals using Walsh and Fourier transforms*, IEEE Trans. Biomed. Engrg., BME-28, pp. 836–838.

P. I. M. JOHANNESMA (1968), *Diffusion models for the stochastic activity for neurons*, in Neural Networks, E. R. Caianiello, ed., Springer-Verlag, Berlin, New York.

T. L. JOHNSON, S. C. WRIGHT, AND A. SEGALL (1979), *Filtering of muscle artifact from the EEG*, IEEE Trans. Biomed. Engrg., BME-26, pp. 556–563.

G. JONA-LASINIO AND P. K. MITTER (1985), *On the stochastic quantization of field theory*, Comm. Math. Phys., 101, pp. 409–436.

D. JUNGE (1981), *Nerve and Muscle Excitation*, Sinauer, Sunderland, MA.

D. JUNGE AND G. P. MOORE (1966), *Interspike-interval fluctuations in Aplysia pacemaker neurons*, Biophys. J., 6, pp. 411–434.

G. KALLIANPUR (1980), *Stochastic Filtering Theory*, Springer-Verlag, Berlin, New York.

—— (1983), *On the diffusion approximation to a discontinuous model for a single neuron*, in Contributions to Statistics, P. K. Sen, ed., North-Holland, Amsterdam.

G. KALLIANPUR AND R. WOLPERT (1984), *Infinite dimensional stochastic differential equation models for spatially distributed neurons*, Appl. Math. Optim., 12, pp. 125–172.

—— (1987), *Weak convergence of stochastic neuronal models*, in Stochastic Methods in Biology, M. Kimura, G. Kallianpur, and T. Hida, eds., Springer-Verlag, Berlin, New York.

D. KANNAN (1979), *An Introduction to Stochastic Processes*, North-Holland, Amsterdam.

B. KATZ AND R. MILEDI (1970), *Membrane noise produced by acetylcholine*, Nature, 226, pp. 962–963.

—— (1972), *The statistical nature of the acetylcholine potential and its molecular components*, J. Physiol., 224, pp. 665–699.

J. KEILSON (1963), *The first passage time density for homogeneous skip-free walks on the continuum*, Ann. Math. Statist., 34, pp. 1003–1011.

J. KEILSON AND N. D. MERMIN (1959), *The second-order distribution of integrated shot noise*, IRE Trans. Inform. Theory, IT-5, pp. 75–77.

J. KEILSON AND H. F. ROSS (1975), *Passage time distributions for Gaussian Markov (Ornstein–Uhlenbeck) statistical processes*, Select. Tables in Math. Statist., 3, pp. 233–327.

C. KOCH, T. POGGIO, AND V. TORRE (1983), *Nonlinear interactions in a dendritic tree: Localization, timing and role in information processing*, Proc. Nat. Acad. Sci. U.S.A., 80, pp. 2799–2802.

J. A. KOZIOL AND H. C. TUCKWELL (1978), *Analysis and estimation of synaptic densities and their spatial variation on the motoneuron surface*, Brain Res., 150, pp. 617–624.

H. I. KRAUSZ (1975), *Identification of nonlinear systems using random impulse train inputs*, Biol. Cybernet., 19, pp. 217–230.

V. I. KRYUKOV (1976), *Wald's identity and random walk models for neuron firing*, Adv. in Appl. Probab., 8, pp. 257–277.

S. W. KUFFLER AND J. G. NICHOLLS (1976), *From Neuron to Brain*, Sinauer, Sunderland, MA.

T. G. KURTZ (1981), *Approximation of Population Processes*, CBMS-NSF Regional Conference Series in Applied Mathematics, 36, Society for Industrial and Applied Mathematics, Philadelphia, PA.

J. W. DE KWAADSTENIET (1982), *Statistical analysis and stochastic modelling of neuronal spike-train activity*, Math. Biosci., 60, pp. 17–71.

H. S. LAM AND D. G. LAMPARD (1981), *Modelling of drug receptor interaction with birth and death processes*, J. Math. Biol., 12, pp. 153–172.

H. LANDAHL (1941), *Theory of the distribution of response times in nerve fibers*, Math. Biophys., 3, pp. 141–147.

H. LANDAHL, W. S. MCCULLOCH, AND W. PITTS (1943), *A statistical consequence of the logical calculus of nervous nets*, Bull. Math. Biophys., 5, pp. 135–137.

P. LÁNSKÝ (1983), *Inference for the diffusion models of neuronal activity*, Math. Biosci., 67, pp. 247–260.

―――― (1984), *On approximations of Stein's neuronal model*. J. Theoret. Biol., 107, pp. 631–647.

P. LÁNSKÝ AND V. LÁNSKÁ (1987), *Diffusion approximations of the neuronal model with synaptic reversal potentials*, Biol. Cybernet., 56, pp. 19–26.

L. LAPICQUE (1907), *Recherches quantitatives sur l'excitation électrique des nerfs traiteé comme une polarization*, J. Physiol. Pathol. Gen., 9, pp. 620–635.

H. LARSEN AND D. C. LAI (1980), *Walsh spectral estimates with applications to the classification of EEG signals*, IEEE Trans. Biomed. Engrg., BME-27, pp. 485–492.

J. F. LAWLESS (1982), *Statistical Models and Methods for Lifetime Data*, John Wiley, New York.

H. LECAR AND R. NOSSAL (1971), *Theory of threshold fluctuations in nerves*, Biophys. J., 11, pp. 1048–1067.

H. LECAR AND F. SACHS (1981), *Membrane noise analysis*, in Excitable Cells in Tissue Culture, P. G. Nelson and M. Lieberman, eds., Plenum Press, New York.

P. A. LEE (1979), *Some stochastic problems in neurophysiology*, Southeast Asian Bull. Math., 11, pp. 205–244.

Y. W. LEE AND M. SCHETZEN (1965), *Measurement of the Wiener kernels of a nonlinear system by cross-correlation*, Internat. J. Control, 2, pp. 237–254.

D. S. LEVINE (1983), *Neural population modeling and psychology: A review*, Math. Biosci., 66, pp. 1–86.

P. A. W. LEWIS (1965), *Some results on tests for Poisson process*, Biometrika, 52, pp. 67–78.

R. S. LIPTSER AND A. N. SHIRYAYEV (1984), *Weak convergence of a sequence of semimartingales to a process of diffusion type*, Math. USSR Sb., 49, pp. 171–195.

I. S. LOSEV (1975), *Model of the impulse activity of a neurone receiving a steady impulse influx*, Biofizika, 20, pp. 893–900.

I. S. LOSEV, M. L. SHIK, AND A. S. YAGODNITSYN (1975), *Method of evaluating the synaptic influx to a single neurone of the mid-brain*, Biofizika, 20, pp. 901–908.

K. L. MAGLEBY AND C. F. STEVENS (1972), *A quantitative description of end-plate currents*, J. Physiol., 223, pp. 173–197.

R. MARCUS (1974), *Parabolic Ito equations*, Trans. Amer. Math. Soc., 198, pp. 177–190.

P. Z. MARMARELIS AND V. Z. MARMARELIS (1978), *Analysis of Physiological Systems: The White Noise Approach*, Plenum Press, New York.

P. Z. MARMARELIS AND K. NAKA (1973), *Nonlinear analysis and synthesis of receptive field responses in the catfish retina*, J. Neurophysiol., 36, pp. 605–618.

G. D. MCCANN (1974), *Nonlinear identification theory models for successive stages of visual nervous systems of flies*, J. Neurophysiol., 37, pp. 869–895.

W. S. MCCULLOCH AND W. PITTS (1943), *A logical calculus of the ideas immanent in nervous activity*, Bull. Math. Biophys., 7, pp. 89–93.

——— (1948), *The statistical organization of nervous activity*, Biometrics, 4, pp. 91–99.

H. P. MCKEAN (1973), *Wiener's theory of nonlinear noise*, in Stochastic Differential Equations, American Mathematical Society, Providence, RI.

D. R. MCNEIL AND S. SCHACH (1973), *Central limit analogues for Markov population processes*, J. Roy. Statist. Soc. Ser. B, 35, pp. 1–23.

R. MILLECHIA AND T. MCINTYRE (1978), *Automatic nerve impulse identification and separation*, Comput. Biomed. Res., 11, pp. 459–468.

M. D. MIYAMOTO (1975), *Binomial analysis of quantal transmitter release of glycerol treated frog neuromuscular junctions*, J. Physiol., 250, pp. 121–142.

C. E. MOLNAR AND R. R. PFEIFFER (1968), *Interpretation of spontaneous spike discharge patterns of neurons in the cochlear nucleus*, Proc. IEEE, 56, pp. 993–1004.

L. E. MOORE AND B. N. CHRISTENSEN (1985), *White noise analysis of cable properties of neuroblastoma cells and lamprey central neurons*, J. Neurophysiol., 53, pp. 636–651.

L. G. MUSCHAWECK AND D. LOEVNER (1978), *Analysis of neuronal spike trains*, Internat. J. Neurosci., 8, pp. 51–60.

J. S. NAGUMO, S. ARIMOTO, AND S. YOSHIZAMA (1962), *An active pulse transmission line simulating nerve axon*, Proc. IRE, 50, pp. 2061–2070.

K. NAKA, M. SAKURANAGA, AND Y. ANDO (1985), *White noise analysis as a tool in visual physiology*, in Contemporary Sensory Neurobiology, M. J. Correia and A. A. Perachio, eds., Walter Liss, New York.

E. NEHER AND B. SAKMANN (1976), *Single channel currents recorded from membrane of denervated frog muscle fibres*, Nature, 260, pp. 799–802.

E. NEHER AND C. F. STEVENS (1977), *Conductance fluctuations and ionic pores in membranes*, Ann. Rev. Biophys. Bioengrg., 6, pp. 345–381.

B. NEUMCKE (1978), *1/f noise in membranes*, Biophys. Struct. Mechanism, 4, pp. 179–199.

H-G. NILSSON (1977), *Estimation of parameters in a diffusion neuron model*, Comput. Biomed. Res., 10, pp. 191–197.

H. OGURA (1972), *Orthogonal functionals of the Poisson process*, IEEE Trans. Inform. Theory, IT-18, pp. 473–481.

A. PAPOULIS (1977), *Signal Analysis*, McGraw-Hill, New York.

G. PARISI AND Y. WU (1981), *Perturbation theory without gauge fixing*, Sci. Sinica, 24, pp. 483–496.

W. J. PARK (1970), *A multi-parameter Gaussian process*, Ann. Math. Statist., 41, pp. 1582–1595.

E. S. PEARSON AND H. O. HARTLEY (1972), *Biometrika Tables for Statisticians, Vol. 2*, Cambridge University Press, Cambridge.

C. PECHER (1939), *La fluctuation d'excitabilité de la fibre nerveuse*, Arch. Internat. Physiol., 49, pp. 129–152.

D. H. PERKEL, G. L. GERSTEIN, AND G. P. MOORE (1967), *Neuronal spike trains and stochastic point processes. II. Simultaneous spike trains*, Biophys. J., 7, pp. 419–440.

R. PYKE (1959), *The supremum and infimum of the Poisson process*, Ann. Math. Statist., 30, pp. 568–576.

W. RALL (1955a), *A statistical theory of monosynaptic input-output relations*, J. Cell. Comp. Comparative Physiol., 46, pp. 373–411.

——— (1955b), *Experimental monosynaptic input-output relations in the mammalian spinal cord*, J. Cell. Comp. Comparative Physiol., 46, pp. 413–437.

——— (1959), *Branching dendritic trees and motoneuron membrane resistivity*, Exper. Neurol., 1, pp. 491–527.

N. RASHEVSKY (1938), *Mathematical Biophysics,* University of Chicago Press, Chicago, IL. (Dover edition, 1960.)

R. REBOLLEDO (1979), *Martingales et convergence étroite de mesures de probabilité,* Kybernetika, 15, pp. 1–7.

S. J. REDMAN, D. G. LAMPARD, AND P. ANNAL (1968), *Monosynaptic stochastic stimulation of cat spinal motoneurons. II. Frequency transfer characteristics of tonically discharging motoneurons,* J. Neurophysiol., 31, pp. 499–508.

L. M. RICCIARDI AND L. SACERDOTE (1979), *The Ornstein–Uhlenbeck process as a model for neuronal activity,* Biol. Cybernet., 35, pp. 1–9.

J. ROBINSON (1976), *Estimation of parameters for a model of transmitter release at synapses,* Biometrics, 32, pp. 61–68.

R. W. RODIECK, N. Y-S. KIANG, AND G. L. GERSTEIN (1962), *Some quantitative methods for the study of spontaneous activity of single neurons,* Biophys. J., 2, pp. 351–368.

A. ROSENBLEUTH, N. WIENER, W. PITTS, AND J. GARCIA RAMOS (1949), *A statistical analysis of synaptic excitation,* J. Cell. Comp. Comparative Physiol., 34, pp. 173–205.

B. ROUX AND R. SAUVE (1985), *A general solution to the time interval omission problem,* Biophys. J., 48, pp. 149–158.

B. K. ROY AND D. R. SMITH (1969), *Analysis of the exponential decay model of the neuron showing frequency threshold effects,* Bull. Math. Biophys., 31, pp. 341–357.

M. SAKURANAGA AND K. NAKA (1985), *Signal transmission in the catfish retina,* J. Neurophysiol., 53, pp. 373–388.

A. C. SANDERSON (1980), *Adaptive filtering of neuronal spike train data,* IEEE Trans. Biomed. Engrg., BME-27, pp. 271–274.

S. SATO (1978), *On the moments of the firing interval of the diffusion approximated neuron,* Math. Biosci., 39, pp. 53–70.

M. SCHETZEN (1980), *The Volterra and Wiener Theories of Nonlinear Systems,* John Wiley, New York.

A. SEGALL AND T. KAILATH (1976), *Orthogonal functionals of independent-increment processes,* IEEE Trans. Inform. Theory, IT-22, pp. 287–298.

G. F. SIMMONS (1963), *Introduction to Topology and Modern Analysis,* McGraw-Hill, New York.

E. SKAUGEN (1978), *The Effects of a Finite Number of Sodium and Potassium Conducting Pores upon Firing Behaviour in Nerve Models,* Institut Informatikk, University of Oslo, Oslo, Norway.

C. E. SMITH AND M. V. SMITH (1984), *Moments of voltage trajectories for Stein's model with synaptic reversal potentials,* J. Theoret. Neurobiol., 3, pp. 67–77.

D. R. SMITH AND G. K. SMITH (1965), *A statistical analysis of the continual activity of single cortical neurones,* Biophys. J., 5, pp. 47–74.

R. B. STEIN (1965), *A theoretical analysis of neuronal variability,* Biophys. J., 5, pp. 173–194.

―――― (1967), *Some models of neuronal variability,* Biophys. J., 7, pp. 37–68.

C. STONE (1963), *Limit theorems for random walks, birth and death processes and diffusion processes,* Illinois J. Math., 7, pp. 638–660.

F. L. H. M. STUMPERS (1950), *On a first-passage time problem,* Philips Res. Rep., 5, pp. 270–281.

H. SUGIYAMA, G. P. MOORE, AND D. H. PERKEL (1970), *Solutions for a stochastic model of neuronal spike production,* Math. Biosci., 8, pp. 323–341.

N. TAKIZAWA AND M. OONUKI (1980), *Analysis of synapse density in cerebral cortex,* J. Theoret. Biol., 82, pp. 573–590.

M. U. THOMAS (1975), *Some mean first passage time approximations for the Ornstein–Uhlenbeck process,* J. Appl. Probab., 12, pp. 600–604.

K. TOYAMA AND K. TANAKA (1984), *Visual cortical functions studied by cross-correlation analysis,* in Dynamic Aspects of Neocortical Function, G. M. Edelman, W. E. Gall, and W. M. Cowan, eds., John Wiley, New York.

J. P. TREMBLOY, R. E. LAURIE, AND M. COLONNIER (1983), *Is the MEPP due to the release of one vesicle or to the simultaneous release of several vesicles at one active zone?* Brain Res. Rev., 6, pp. 299–314.

A. TSURUI AND S. OSAKI (1976), *On a first-passage problem for a cumulative process with exponential decay,* Stochastic Proc. Appl., 4, pp. 79–88.

H. C. TUCKWELL (1974), *A study of some diffusion models of population growth*, Theor. Pop. Biol., 5, pp. 345–357.

────── (1975), *Determination of the inter-spike times of neurons receiving randomly arriving post-synaptic potentials,* Biol. Cybernet., 18, pp. 225–237.

────── (1976a), *On the first-exit time problem for temporally homogeneous Markov processes,* J. Appl. Probab., 13, pp. 39–48.

────── (1976b), *Frequency of firing of Stein's model neuron with application to cells of the dorsal spinocerebellar tract,* Brain Res., 116, pp. 323–328.

────── (1976c), *Firing rates of motoneurons with strong random synaptic excitation,* Biol. Cybernet., 24, pp. 147–152.

────── (1978a), *Recurrent inhibition and afterhyperpolarization: Effects on neuronal discharge,* Biol. Cybernet., 30, pp. 115–123.

────── (1978b), *Neuronal interspike time histograms for a random input model,* Biophys. J., 21, pp. 289–290.

────── (1979), *Synaptic transmission in a model for stochastic neural activity,* J. Theoret. Biol., 77, pp. 65–81.

────── (1981), *Poisson processes in biology,* in Stochastic Nonlinear Systems, Springer-Verlag, Berlin, New York, pp. 162–171.

────── (1985), *Some aspects of cable theory with synaptic reversal potentials,* J. Theoret. Neurobiol., 4, pp. 113–127.

────── (1986a), *On shunting inhibition,* Biol. Cybernet., 55, pp. 83–90.

────── (1986b), *Stochastic equations for nerve membrane potential,* J. Theoret. Neurobiol., 5, pp. 87–99.

────── (1987a), *Statistical properties of perturbative nonlinear random diffusion from stochastic integral representations,* Phys. Lett. A, 122, pp. 117–120.

────── (1987b), *Diffusion approximations to channel noise,* J. Theoret. Biol., 127, pp. 427–438.

────── (1988a), *Introduction to Theoretical Neurobiology, Volume 1: Linear Cable Theory and Dendritic Structure,* Cambridge University Press, New York.

────── (1988b), *Introduction to Theoretical Neurobiology, Volume 2: Nonlinear and Stochastic Theories,* Cambridge University Press, New York.

────── (1988c), *Perturbative analysis of random nonlinear reaction-diffusion systems,* Phys. Scripta, 37, pp. 321–322.

────── (1988d), *On the use of Green's function matrices for systems of diffusion equations,* Internat. J. Systems Sci., 19, pp. 1663–1666.

H. C. TUCKWELL AND D. K. COPE (1980), *Accuracy of neuronal interspike times calculated from a diffusion approximation,* J. Theoret. Biol., 83, pp. 377–387.

H. C. TUCKWELL AND W. RICHTER (1978), *Neuronal interspike time distributions and the estimation of neurophysiological and neuroanatomical parameters,* J. Theoret. Biol., 71, pp. 167–183.

H. C. TUCKWELL AND J. B. WALSH (1983), *Random currents through nerve membranes,* Biol. Cybernet., 49, pp. 99–110.

H. C. TUCKWELL AND F. Y. M. WAN (1980), *The response of a nerve cylinder to spatially distributed white noise inputs,* J. Theoret. Biol., 87, pp. 275–295.

────── (1984), *First passage time of Markov processes to moving barriers,* J. Appl. Probab., 21, pp. 695–709.

H. C. TUCKWELL, F. Y. M. WAN, AND Y. S. WONG (1984), *The interspike interval of a cable model neuron with white noise input,* Biol. Cybernet., 49, pp. 155–167.

G. E. UHLENBECK AND L. S. ORNSTEIN (1930), *On the theory of Brownian motion,* Phys. Rev., 36, pp. 823–841.

A. M. UTTLEY (1956), *The probability of neural connexions,* Proc. Roy. Soc. Ser. B., 144, pp. 229–240.

W. VAN DER KLOOT, H. KITA, AND I. COHEN (1975), *The timing and appearance of miniature end-plate potentials,* Progr. Neurobiol., 4, pp. 271–326.

G. VENZL (1976), *Statistical fluctuations of activity in localized neural populations,* J. Theoret. Biol., 63, pp. 259–274.

D. VERE-JONES (1966), *Simple stochastic models for the release of quanta of transmitter from a nerve terminal*, Australian J. Statist., 8, pp. 53–63.

A. A. VERVEEN AND L. J. DE FELICE (1974), *Membrane noise*, Progr. Biophys. Molec. Biol., 28, pp. 189–265.

A. A. VERVEEN AND H. E. DERKSEN (1965), *Fluctuations in membrane potential of axons and the problem of coding*, Kybernetik, 4, pp. 152–160.

L. WALLOE, J. K. S. JANSEN, AND K. NYGAARD (1969), *A computer simulated model of a second order sensory neurone*, Kybernetik 6, pp. 130–141.

R. E. WALPOLE AND R. H. MYERS (1985), *Probability and Statistics for Engineers and Scientists*, Macmillan, New York.

J. B. WALSH (1981), *A stochastic model of neuronal response*, Adv. Appl. Prob., 13, pp. 231–281.

J. B. WALSH AND H. C. TUCKWELL (1985), *Determination of the electrical potential over dendritic trees by mapping onto a nerve cylinder*, J. Theoret. Neurobiol, 4, pp. 27–46.

D. O. WALTER (1969), *A posteriori "Wiener filtering" of average evoked responses*, EEG Clinical Neurophysiol. Suppl., 27, pp. 61–70.

F. Y. M. WAN (1972), *Linear partial differential equations with random forcing*, Stud. Appl. Math., 51, pp. 163–178.

F. Y. M. WAN AND H. C. TUCKWELL (1979), *The response of a spatially distributed neuron to white noise current injection*, Biol. Cybernet., 33, pp. 39–55.

—— (1982), *Neuronal firing and input variability*, J. Theoret. Neurobiol., 1, pp. 197–218.

M. C. WANG AND G. E. UHLENBECK (1945), *On the theory of Brownian motion II*, Rev. Modern Phys., 17, pp. 323–342.

B. S. WHITE AND S. ELLIAS (1979), *A stochastic model for neuronal spike generation*, SIAM J. Appl. Math., 37, pp. 206–233.

N. WIENER (1949), *Extrapolation, Interpolation and Smoothing of Stationary Time Series*, John Wiley, New York.

—— (1958), *Nonlinear Problems in Random Theory*, John Wiley, New York.

W. J. WILBUR AND J. RINZEL (1982), *An analysis of Stein's model for stochastic neuronal excitation*, Biol. Cybernet., 45, pp. 107–114.

—— (1983), *A theoretical basis for large coefficient of variation and bimodality in neuronal interspike interval distributions*, J. Theoret. Biol., 105, pp. 345–368.

H. R. WILSON AND J. D. COWAN (1972), *Excitatory and inhibitory interactions in localized populations of model neurons*, Biophys. J., 12, pp. 1–27.

—— (1973), *A mathematical theory of the functional dynamics of cortical and thalamic nervous tissue*, Kybernetik, 13, pp. 55–80.

E. WONG (1971), *Stochastic Processes in Information and Dynamical Systems*, McGraw-Hill, New York.

E. WONG AND M. ZAKAI (1974), *Martingales and stochastic integrals for processes with a multi-dimensional parameter*, Z. Wahrschein Verw. Gebiete, 29, pp. 109–122.

A. M. YAGLOM (1962), *An Introduction to the Theory of Stationary Random Functions*, Dover, New York.

K. YANA, N. TAKEUCHI, Y. TAKIKAWA, AND M. SHIMOMURA (1984), *A method for testing an extended Poisson hypothesis of spontaneous quantal transmitter release at neuromuscular junctions*, Biophys. J., 46, pp. 323–330.

G. L. YANG AND T. C. CHEN (1978), *On statistical methods in spike train analysis*, Math. Biosci., 38, pp. 1–34.

S. YASUI (1979), *Stochastic functional Fourier series, Volterra series and nonlinear systems analysis*, IEEE Trans. Automat. Control., AC-24, pp. 230–242.

G. F. YEO (1974), *A finite dam with exponential release*, J. Appl. Probab., 11, pp. 122–133.

G. J. ZIMMERMAN (1972), *Some sample function properties of the two-parameter Gaussian process*, Ann. Math. Statist., 43, pp. 1235–1246.

Index

action potential, 1, 3, 4, 7, 8, 17, 18, 20, 21, 24, 29, 38, 48, 66
afterhyperpolarization, 18, 48
autocorrelation coefficients, 87
axons 3, 16

birth and death process, 24, 47, 96–98
boundary classification, 46
boundary conditions, 17, 35, 36, 37, 46, 50, 63

cable equation, 6, 7, 17
cable model, 58
   neuron, 68
Cameron and Martin's expansion, 103, 105
channel noise, see noise
coefficient of variation, 45
conditional means test, 87
cross-intensity function, 89

dendrites, 1, 3, 6, 7, 57–58, 60
dendritic geometry, 59
depolarization, 3, 6, 17, 20, 23, 29–30, 40, 48, 58, 62, 66
differential-difference equation, 35–37, 41
diffusion approximations, 97, 99
diffusion processes, 43, 55

electroencephalograph (EEG), 2, 8, 111–114
   records, 114
end-plate potential (EEP), 11–12, 95
   miniature (MEPP), 9, 11–12
evoked potentials (EPs), 2, 8, 22, 111–112
excitatory postsynaptic potentials (EPSP), see postsynaptic potentials
expectation density, 82

filtering, 89
first exit time, 34, 37
   moments of, 34

first passage time, 18, 22–23, 25, 27, 35, 37–39, 45, 47, 49–51, 55, 66, 71, 75
Fitzhugh–Nagumo equations (stochastic) 8, 73, 78, 79
   system of, 71, 76–77
fluctuation analysis, 92
Fourier transform, 33
Fourier–Hermite set, 104–105

Goldman–Hodgkin–Katz formula, 4
goodness-of-fit test, 12, 84
Green's function, 7, 32, 58, 61, 101, 62–65, 67, 68, 78
   matrix of, 79

Hilbert space, 102
historical review, 1–2, 11, 15
Hodgkin–Huxley equations (stochastic), 1, 8, 58, 71–74, 79, 80

independent increments, 25
infinite cable, 64
infinitesimal generator, 31, 40
inhibitory postsynaptic potentials (IPSPs), see postsynaptic potentials
interspike interval (ISI), 9, 26, 35, 37, 45–46, 47, 49, 68
   classification of their distributions, 84
inverse Gaussian distribution, 45–46

Kolmogorov equation
   backward, 31, 44, 74, 76
   forward, 31, 44, 74
   steady-state, 98

Lapicque model 1, 5, 7, 17
linear cable theory, 1, 5
linear decay, 21

INDEX

Lorentzians, 94–95

mappings to cylinders, 6
Markov chains, 2, 92
Markov processes 2, 30, 34, 43, 57, 62, 66, 67, 74
   discontinuous, 29, 63
maximum likelihood, 83
McCulloch and Pitts models, 16
mean firing time, 36
method of moments, 83
motoneuron firing patterns, 47
multidimensional diffusion, 74

Nernst potentials, 5, 8, 58, 60, 93, 99
neuroanatomy
   basic, 2–3
   stochastic, 114
neuronal activity, 25, 32, 49, 54
neuronal models, 34, 57
   early, 15
   stochastic, 22
neurons
   populations of, 115
   properties of, 1, 2, 3, 15, 16, 17, 19, 26, 30, 35–38, 43, 49, 54, 114–115
noise
   channel, 9, 13, 91–99
   1/f, 9, 91
   Poisson, 58
nonlinear equations, 71
   differential, 1

Ornstein–Uhlenbeck Process (OUP), 2, 5, 43, 46, 47, 48, 49–53, 55, 61, 65, 66, 68, 70, 97

parameter estimation, 88
perfect integrator, 18, 23
periodic random excitation, 22
perturbative analysis, 77
point process, 9, 18, 19
Poisson
   case, 36, 63, 75
   excitation, 20–21
   hypothesis, 12, 21
   input, 39
   point process, 9
   processes, 13, 15, 16, 19, 20, 23, 29, 57, 62, 68, 73, 88, 110
   random measure, 30–31, 75
   white noise, 67
poststimulus time histograms, 81
postsynaptic potentials, 29, 32, 36, 48, 56, 111, 114

excitatory (EPSP), 4, 15, 36, 37, 47, 111
inhibitory (IPSP), 47, 111
probalons, 73

quantum field theory, 78
quantum hypothesis, 11–12

R-C circuit model, 1, 17, 29, 58, 60
Rall model, 60, 71
random excitation (periodic), 22
random field, 59
random graph theory, 115
random walks, 15, 22
   randomized, 23, 29, 44, 55
reaction-diffusion systems, 7
renewal equation, 23
renewal process, 82
reversal potentials, 29, 39, 40–41, 43, 53, 55, 62
runs test, 86

sealed end conditions, 17
separation problems, 89
simulation, 67
Skorohod metric, 55
soma, 3, 7, 58, 60, 62
space-clamped systems, 74–77
   Hodgkin–Huxley, 74, 75
spectral density, 70, 78, 94, 96
spike density, 82
spike rate function, 82
stability property, 24
stationarity, 86
stationary density, 98
stationary processes, 20, 48, 49, 70, 97, 112–113
   distributions of, 99
steady-state distributions, 33, 51, 52, 99
Stein's model, 29, 35, 36, 43, 46, 55, 63, 67
stochastic differential equations, 30, 33, 38, 39–40, 43, 44, 46, 48, 67, 71, 75
stochastic partial differential equations (SPDEs) 32, 57, 58, 60, 62, 65–66, 69, 70, 71, 77, 79
stochastic phenomena, 8
stochastic processes, 1–2
Stratonovich integral, 108
superposition of processes, 19
synapse densities, 4
synapses, 3, 48
synaptic
   currents, 53, 62, 68
   input, 18, 20, 29, 60, 62–63, 68
   reversal potentials, 39
   transmission, 11–12, 32
system identification, 9, 101

tails, 24
threshold, 52, 55, 66
   assumption, 7, 18
   crossings, 32
   effects, 17
   formulas, 37
   properties, 8, 45, 48
   time-varying, 38, 52
trigger zone, 66

uniform metric, 97

voltage, 32, 48, 61, 69
   clamp, 93, 95
   fluctuations, 47
   noise, 99
   variable, 8

weak convergence, 19, 54, 97
white noise
   current, 60, 75, 101, 109
   Gaussian, 72, 74, 76, 77
   Poisson, 58, 77
   potential, 71
   two-parameter, 68–69, 79
Wiener
   expansion, 13, 102
   filtering, 112
   integrals, 61, 101, 102–108
   kernel expansion, 2, 9, 105, 108–109
   processes, 2, 30–31, 43–44, 46, 52, 70, 76, 101, 103, 110, 115
   two-parameter, 69